IMPEDANCE

No. 829
$9.95

IMPEDANCE

By Rufus P. Turner

TAB BOOKS
Blue Ridge Summit, Pa. 17214

FIRST EDITION

FIRST PRINTING—APRIL 1976

Printed in the United States
of America

Hardbound Edition: International Standard Book No. 0-8306-6829-2

Paperbound Edition: International Standard Book No. 0-8306-5829-7

Library of Congress Card Number: 75-41733

Cover photo courtesy of Electronic Technician/Dealer.

Preface

Impedance is an important property of all AC circuits and of many electrical devices. This property is encountered and must be dealt with wherever a signal or power is handled or processed; and the technician who has a good understanding of impedance is at home among many of the complexities of electronics.

From a largely practical point of view, this book surveys the subject of impedance—its nature, how it is calculated, and how it is measured. And because this is a practical book, every effort has been put forth to keep such theroetical discussion as is necessary in such form as to be understandable to the average technician. No mathematical background beyond the leading facts of algebra, trigonometry, and vectors is required, and examples are used generously to reinforce the discussion.

The purpose of the book is to impart a good working knowledge of the subject and also to provide a ready reference for the technician or student when he needs a quick refresher on some aspects of impedance. Obviously, there is much that we have been unable to include, but this book should brace the reader for a subsequent study of more advanced texts.

Rufus P. Turner

Contents

1 AC Fundamentals

This chapter reviews briefly those selected fundamentals of alternating-current electricity that are essential to the understanding of impedance. This is done with the single aim of aiding the reader; hence, the chapter should serve as an introduction to the subject or as a refresher, whichever is needed. The presentation moves from a simple description of alternating current and voltage to a description of alternating currents in reactances—reactance being the logical bridge to impedance.

1.1 NATURE OF ALTERNATING CURRENT

Whereas a direct current (DC) is unidirectional—even when sometimes it rises and falls periodically (pulsating DC)—an alternating current (AC) periodically changes its direction. An alternating current starts at zero, increases to a maximum positive value, decreases through zero to a maximum negative value, and returns to zero. This single, complete set of changes is termed a *cycle*. The cycle is repeated for as long as current flows.

A plot of instantaneous values of current against time shows how the current varies in a particular AC cycle; the shape of this cycle (the *waveshape* or *waveform*) depends

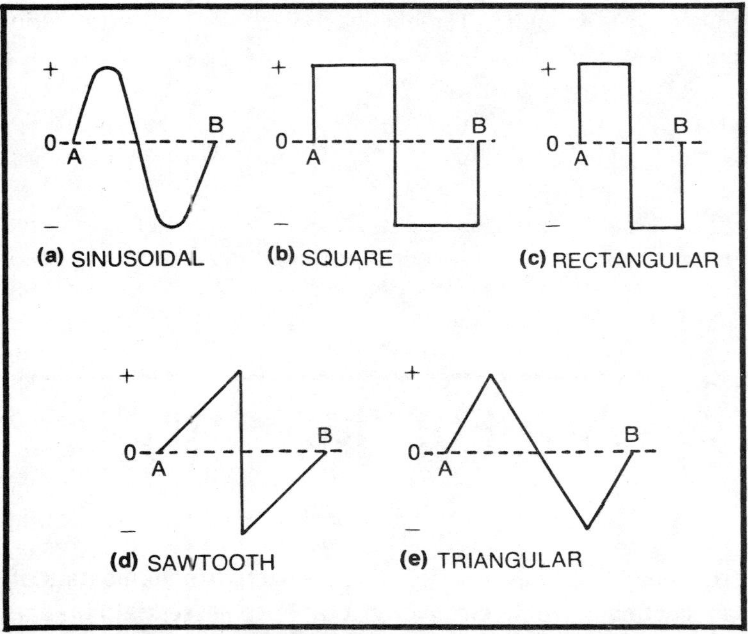

Fig. 1-1. Representative AC waveforms. The period is the duration from point **a** to point **b** in any of these waveshapes.

upon how the current is generated or processed. Common examples are shown in Fig. 1-1. Here Fig. 1-1(a) shows a sine wave; in this cycle, the changes are gradual. In Fig. 1-1(b), however, the current rises abruptly to maximum positive, holds for an interval, drops abruptly through zero and reaches maximum negative, holds for an interval, and finally rises abruptly to zero (this is a square wave). The rectangular wave in Fig. 1-1(c) is similar to the square wave, except that the rectangular wave holds at positive maximum and negative maximum for shorter intervals than the square wave. The sawtooth wave in Fig. 1-1(d) is characterized by a slow, usually linear increase from zero to maximum positive and a similar change from maximum negative back to zero, but with an abrupt intermediate change from maximum positive to maximum negative. By contrast, the triangular wave in Fig. 1-1(e) has a similar angular climb from zero to positive maximum and from negative maximum to zero, but an angular, rather than abrupt, change from positive maximum to negative maximum. Each of these waveshapes has specific applications in electronics.

There can be, and very often are, AC cycles having shapes other than those shown in Fig. 1-1. The cycles shown in this illustration are symmetrical; that is, the positive half-cycle is of the same size and shape as the negative half-cycle. But waveshapes that are asymmetrical—either vertically or horizontally, or both—are sometimes encountered. These latter waveshapes are said to be *complex*. While Fig. 1-1 shows only cycles that go first to positive maximum and then to negative maximum, the opposite state—going first to negative maximum—also exists.

Alternating voltage and current are associated in the same sense that direct current and voltage are associated. Accordingly, alternating current may be thought of in terms of being produced by AC voltage, and the flow of alternating current through a resistor is seen to set up an AC voltage drop across that resistor. The alternating voltage cycle resembles the alternating current cycle, and vice versa; because of distortion, though, the two might not always be *exact* replicas.

1.2 FREQUENCY

The term *frequency* (f) denotes the number of complete cycles occurring in one second—the number of cycles per second, or *hertz*; thus, hertz is the basic unit of frequency.

The hertz is not always a practically manageable unit; many of the frequencies regularly employed in electronics are extremely high by comparison. In microwave practice, frequencies often are in excess of 10 billion hertz. Larger units than the hertz therefore are required for practical use; these are kilohertz (*kHz*), megahertz (*MHz*), and gigahertz (*GHz*). The prefixes kilo, mega, and giga stand for thousand, million, and billion. Table 1-1 lists common frequency units, and Table 1-2 shows how to convert from one unit to another.

Example 1.1. The frequency of Citizens Band channel 9 is 27.065 MHz. What does this correspond to in kilohertz?

From Table 1-2, 1 MHz = 10^6 Hz, or 10^3 kHz. So,

$$f = 27.065 \times 1000 = 27065 \text{ kHz}.$$

Table 1-1. Common Frequency Units.

1 cps = 1 Hz
1 kHz = 1000 Hz
1 MHz = 1,000,000 Hz
1 GHz = 1,000,000,000 Hz

Frequency is an important quantity in impedance calculations and measurements, since impedance is a frequency-dependent property.

1.3 PERIOD

The term *period* (t) denotes the total time it takes for a voltage or current to complete one full cycle. This is the distance from *a* to *b* in any of the cycles of Fig. 1-1. Obviously, the higher the frequency, the more cycles occurring in one second, and the shorter the period of each cycle. Period has a simple relationship to frequency:

$$t = 1/f \qquad (1\text{-}1)$$

where t is in seconds and f is in hertz.

Example 1-2. Calculate the period of a 2 kHz signal.

From Eq. 1-1, 2 kHz = 2000 Hz. From Eq. 1-1,

$$t = 1/2000 = 0.0005 \text{ second.}$$

Equation 1-1 and the example give time in seconds. In practice, however, one second is often a long interval and subdivisions of this unit must be used: milliseconds (thousandths of a second, abbreviated *ms* or *msec*), microseconds (millionths of a second, abbreviated *μs* or *μsec*), and nanoseconds (billionths of a second, abbreviated *ns* or *nsec*). Table 1-3 gives the periods of some common frequencies often employed in impedance measurements.

Table 1-2. Frequency Conversion Factors.

Hz	$= 10^{-3}$ kHz	$= 10^{-6}$ MHz	$= 10^{-9}$ GHz
kHz	$= 10^{3}$ Hz	$= 10^{-3}$ MHz	$= 10^{-6}$ GHz
MHz	$= 10^{6}$ Hz	$= 10^{3}$ kHz	$= 10^{-3}$ GHz
GHz	$= 10^{9}$ Hz	$= 10^{6}$ kHz	$= 10^{3}$ MHz

These periods are given in the time units used most often with the frequencies noted.

1.4 SINE WAVE

The earliest source of useful amounts of AC power was a rotating machine—a *generator* in which a coil rotating in the uniform field between the two poles of a magnet has a voltage induced across it. Simplified for purposes of explanation, the coil could consist of a single loop of wire. Across such a coil turning in an imaginary circle, the induced voltage increases from zero to maximum positive and returns to zero as one side of the coil moves past one pole; then the voltage "increases" from zero to maximum negative and returns to zero as the same side of the coil moves past the opposite pole. Thus, in 360 degrees of coil rotation (one complete revolution), the voltage describes the AC cycle: zero, positive maximum, zero, negative maximum, zero. This pattern is illustrated in Fig. 1-1.

At any instant, the corresponding voltage is proportional to the sine of the angle through which the coil has turned, and this is responsible for the characteristic waveshape (Fig. 1-1) resulting from this action and for the term *sine wave*. This, of

Table 1-3. Values of Period for Common Frequencies.

f	t
20 Hz	50 ms
30 Hz	33.3 ms
40 Hz	25 ms
50 Hz	20 ms
60 Hz	16.7 ms
100 Hz	10 ms
120 Hz	8.3 ms
400 Hz	2.5 ms
500 Hz	2.0 ms
1000 Hz	0.1 ms
2500 Hz	400 μs
10 kHz	100 μs
20 kHz	50 μs
100 kHz	10 μs
1 MHz	1.0 μs

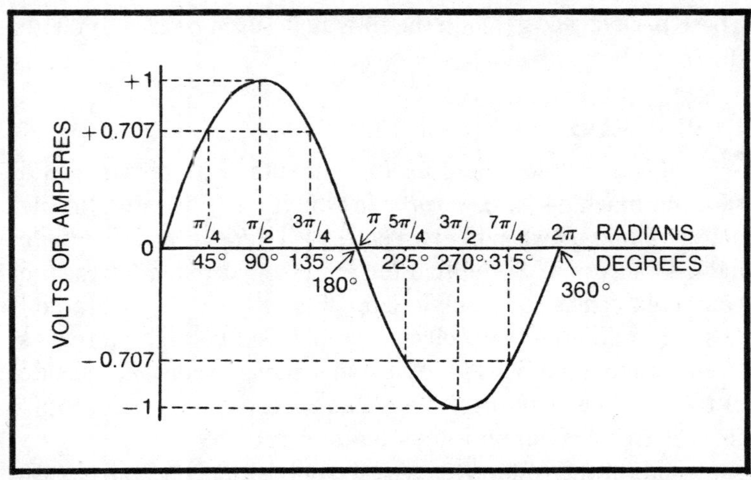

Fig. 1-2. A single sine-wave cycle with voltage plotted against the angle of rotation in both degrees and radians.

course, is the curve of the sine function in trigonometry. The sine wave has great utility in electronics. Other waves—a few examples of which appear in Fig. 1-1 —are called *nonsinusoidal*. To find the instantaneous voltage (e) at any angle (θ) in the rotation of the coil, it is necessary only to multiply the maximum value the voltage will attain (E_{MAX}) by the sine of that angle:

$$e = E_{MAX} \sin \theta \qquad (1\text{-}2)$$

where e and E_{MAX} are in the same units (V, mV, μV).

Example 1-3. The maximum voltage (positive or negative) reached by a certain sine wave is 6.3V. Calculate the instantaneous voltage at 60 degrees.

The sine of 60 degrees is 0.866025. From Eq. 1-2,

$$e = 6.3(0.866025)$$
$$= 5.45\text{V}$$

Figure 1-2 shows a single sine-wave cycle with voltage plotted against the angle of rotation in both degrees and radians. If, as in this sketch, a maximum value of 1V is assumed, the voltage at the instant when the angle is 45 degrees ($\pi/4$ radians) is 0.707V, since sin 45 degrees = 0.707, and the instantaneous voltage (from Eq. 1-2) is $1 \times 0.707 = 0.707V$. Note that the instantaneous voltage is again 0.707V at 135 degrees, since sin 135 degrees = 0.707.

Generators still produce most of our electrical energy, but they are seldom found in electronic equipment. A high-grade oscillator employing transistors or tubes also generates a sine wave and has no moving parts. Nevertheless, the angles (which originally denoted positions of the moving coil in a machine) apply to the oscillator signal as well, and must be used in many AC calculations. In modern practice, however, it is often more convenient to plot the AC cycle on a horizontal time axis (as when the signal is presented on an oscilloscope screen) and to convert the time units to corresponding angles. In this connection, Fig. 1-3 shows a single cycle of a 1000 Hz sine wave. Note that the period here is one millisecond (refer to Sec. 1.3) and that the instantaneous voltage at several intermediate instants is noted: 0.125, 0.25, 0.375, 0.5, 0.625, 0.75, 0.875, and 1 ms. At any instant t, the angle θ may be calculated in terms of frequency and time:

$$\theta = 2\pi f t \qquad (1\text{-}3)$$

where θ is in radians, f in hertz, and t in seconds.

Example 1-4. Calculate the angle in degrees at the 0.125 ms point in the 1000 Hz cycle shown in Fig. 1-3.

Here, 0.125 ms = 0.000125s. From Eq. 1-3,

$$\begin{aligned} \theta &= 2(3.1416)1000(0.000125) \\ &= 6.2832(0.125) \\ &= 0.7854 \text{ radian} \\ 45 &= \text{degrees.*} \end{aligned}$$

Example 1-5. Calculate the angle in degrees at the 0.75 ms point in the 1000 Hz cycle shown in Fig. 1-3.

Here, 0.75 ms = 0.00075s. From Eq. 1-3,

$$\begin{aligned} \theta &= 2(3.1416)1000(0.00075) \\ &= 6.2832(0.75) \\ &= 4.7122 \text{ rad} \\ &= 270 \text{ degrees.} \end{aligned}$$

---

*Degrees = radians × 57.295. Radians = degrees × 0.0174533.

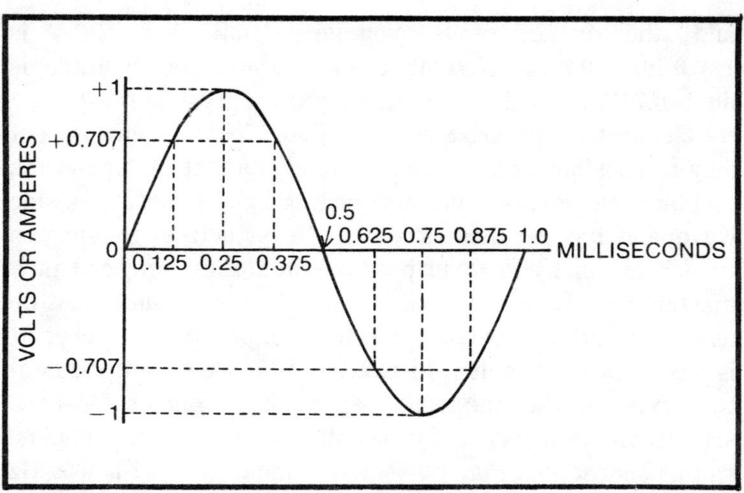

Fig. 1-3. A single-cycle 1000 Hz sine wave with the time axis plotted and several instantaneous voltages noted.

Observe that the instantaneous voltages in Fig. 1-3 are identical with those in Fig. 1-2: e at 0.125 ms and 0.375 ms (corresponding to 45 and 135 degrees, respectively) is +0.707V, and at 0.625 and 0.875 ms (corresponding to 135 and 315 degrees, respectively) is −0.707V. This shows that Eq. 1-2 may be rewritten to give voltage in terms of time:

$$e = E_{MAX} \sin 2\pi ft \qquad (1\text{-}4)$$

where e and E_{MAX} are in the same units (V, mV, μV), f is in hertz, and t in seconds.

Thus, from Eq. 1-4, the instantaneous voltage at 0.75 ms is equal to $E_{MAX} \sin [2(3.1416)1000(0.00075)] = E_{MAX} \sin 4.7124$ rad $= E_{MAX} \sin 270$ degrees $= -1(1) = -1$V. The instantaneous voltage may be found in this way for any instant in a cycle of any frequency. From this discussion, it should be clear that the expression $2\pi ft$ equals the angle in radians.

The quantity $2\pi f$ in Eq. 1-4 is often encountered in engineering formulas and is frequently abbreviated by the lowercase Greek omega (ω). This changes Eq. 1-4 to:

$$e = E_{MAX} \sin \omega t \qquad (1\text{-}5)$$

1.5 ANGULAR VELOCITY

The symbol ω, which appears first in Eq. 1-5 and is equal to $2\pi f$, is the symbol for *angular velocity*. This symbol appears in a great many AC formulas.

To grasp the physical significance of angular velocity in this sense, we must return to the mechanical AC generator. In this machine, the conductor rotates through an angle of 2π radians during each revolution, since there are 2π radians in a circle, and the angular velocity of the rotating shaft is thus the product of 2π radians times the number of revolutions per second. The equivalent electrical quantity is the product of 2π radians and the AC frequency (cycles per second, or hertz, replacing revolutions per second, since one electrical cycle is equivalent to one mechanical revolution). As in the mechanical example, this is also expressed in radians per second. Thus, for 400 Hz: $\omega = 2\pi f = 2(3.1416)400 = 2513$ radians per second.

Table 1-4 lists values of ω for 23 common frequencies between 20 Hz and 1 MHz.

1.6 AC COMPONENTS AND VALUES

In its 360-degree (2π radians) excursion, the AC cycle passes through a number of voltage or current values. Which of these is significant depends upon the nature of the application or calculation involved. The four terms which describe the AC component are *maximum value*, *instantaneous value*, *average value*, and RMS *value*.

Maximum Value

This is the highest positive or negative value reached in the cycle. It is also called *peak value*. It is the value to which a peak-responding electronic voltmeter (such as the rectifier/amplifier type) responds, and it is also the value which determines the no-load output of voltage doublers, triplers, and quadruplers. Many electronic circuits are adjusted on the basis of the maximum value of the AC signal.

Instantaneous Value

This is the value at any selected instant during the cycle. Instantaneous voltage or current is sometimes labeled to show

Table 1-4. Values of Angular Velocity τ for 23 Common Frequencies.

f	ω
20 Hz	125.7
30 Hz	188.5
40 Hz	251.3
50 Hz	314.1
60 Hz	377
100 Hz	628.3
120 Hz	754
150 Hz	942.5
200 Hz	1256
300 Hz	1885
400 Hz	2513
500 Hz	3142
1000 Hz	6283
1500 Hz	9425
2000 Hz	12,566
5000 Hz	31,416
10 kHz	62,832
20 kHz	125,664
50 kHz	314,159
100 kHz	628,318
200 kHz	1,256,637
500 kHz	3,141,592
1 MHz	6,283,185

its exact point along the horizontal axis, thus: $e_{10°}$, $e_{\pi\ 2}$, i_{2ms}, etc. For a sine wave, $e = E_{MAX} \sin \theta$, and $i = I_{MAX} \sin \theta$.

Average Value

This is the simple average (arithmetic mean) of all the instantaneous values in one cycle, disregarding sign. $E_{AVG} = 0.637\ E_{MAX}$, and $I_{AVG} = 0.637\ I_{MAX}$. The larger the number of instantaneous values that enter into the calculation, the more exact the calculation will be. However, without calculus, a phenomenal number of instantaneous values must be used to obtain the number 0.637. The average value is the voltage to which amplifier/rectifier-type electronic voltmeters respond. It is also the value of voltage delivered by an unfiltered full-wave rectifier.

RMS Value

This is the *root mean square* value. It is also called the *effective* value, since it is equivalent to the same-numbered DC value in the heating effect it creates in a resistance. One rms ampere produces the same average heating effect that one DC ampere does:

$$E_{RMS} = 0.707\,E_{MAX}, \text{ and } I_{RMS} = 0.707\,I_{MAX}.$$

The rms value, as its name implies, is equal to the square root of the mean of the squares of all the instantaneous values in one cycle, disregarding sign. To calculate the rms value: square each instantaneous value, but do not include the maximum value; total these squares; take the average (arithmetic mean) of this total; extract the square root of this average. Without calculus, a phenomenal number of instantaneous values must be used to obtain the number 0.707, which we are free to use without first deriving it.

The rms value is the one in which most AC voltmeters and ammeters read, whether or not they actually respond to this value. The widely used rectifier-type meter, for example, is *average*-responsive, but its scale reads in the more useful rms units.

Conversions

Table 1-5 gives multipliers for converting maximum, average, and rms values. The use of these conversion factors is straightforward. To convert $12.6V_{RMS}$ to average volts, multiply by 1.11:

$$12.6 \times 1.11 = 13.99V_{AVG}$$

Table 1-5. Voltage and Current Conversions and RMS Values.

$$E_{AVG} = 0.637\,E_{MAX} = 0.901\,E_{RMS}$$
$$E_{RMS} = 0.707\,E_{MAX} = 1.11\;\;E_{AVG}$$
$$E_{MAX} = 1.414\,E_{RMS} = 1.57\;\;E_{AVG}$$

$$I_{AVG} = 0.637\,I_{MAX} = 0.901\,I_{RMS}$$
$$I_{RMS} = 0.707\,I_{MAX}1.11\;\;I_{AVG}$$
$$I_{MAX} = 1.414\,I_{RMS} = 1.57\;\;I_{AVG}$$

The numbers given in this table and earlier in this section apply to sine-wave voltages and currents only. The relationships are quite different with other waveforms. For instance, in a square wave, $E_{RMS} = E_{AVG} = E_{MAX}$. In a positive-going sawtooth wave, $E_{AVG} = 0.5E_{MAX}$, and $E_{RMS} = 0.577\ E_{MAX}$. This points up the error possible when instruments calibrated with a sine wave are used to check nonsinusoidal current or voltage. The readings of a nonpeak-reading electronic voltmeter equipped with an rms scale can be considerably in error if used to measure square-wave voltage, for example. Likewise, when a sinusoidal quantity under measurement contains harmonics, the error in measurement could equal that of the harmonic percentage. (Sine waves are not multiples of any frequency; presence of harmonics indicates that the wave is *not* actually sinusoidal—distortion is thus present.)

1.7 DISTORTION AND HARMONICS

In an ideal sine wave, the instantaneous voltage at any point is proportional to the sine of the corresponding angle, and the smooth curve of Fig. 1-1(a) results. Such perfection is unattainable in practice; some variation, however minute, occurs in signals from even the most refined sources. A signal that departs from the ideal is termed *distorted*.

A byproduct of distortion, which is at once also the nature of the distortion, is the presence of harmonics. These are extra frequencies which are exact multiples, even or odd, of the main frequency which is called the *fundamental frequency* (f). The fundamental frequency is regarded as the first harmonic, and the others are identified as h_2 (2 times f), h_3 (3 times f), etc. to show whether they are the second harmonic, third harmonic, etc. In most instances distortion is considered a defect, since it wastes energy, creates discord (as in an audio amplifier), and causes errors in impedance measurements. In a few instances, it serves a useful purpose—as in harmonic generators, generators of nonsinusoidal waveforms, and most electronic musical instruments.

Harmonic distortion is evaluated in terms of the relative strengths of harmonic and fundamental components. When a wave analyzer is used to measure these components it is tuned successively to the fundamental frequency and to each of the harmonic frequencies, and the voltage amplitude of each of these components is read from the indicating meter. From this data, the strength of each harmonic may be expressed as a percentage of the strength of the fundamental. Thus, the second harmonic content would be equal to f/h_2, expressed as a percentage. The *total harmonic distortion* (the combined distortion due to all harmonics present) would be:

$$D\% = 100\sqrt{h_2^2 + h_3^2 + h_4^2 + \ldots h_N^2} \qquad (1\text{-}6)$$

The 100 in the equation converts the resulting figure to a percentage. When a distortion meter is used, the combined voltage E_T due to the fundamental frequency *and* its harmonics is first measured. Then the fundamental frequency is removed by means of a high-Q filter, and the remaining voltage (E_H), which is due to harmonics alone, is measured. The total distortion then is calculated:

$$D\% = (100\, E_H)/E_T \qquad (1\text{-}7)$$

Professional distortion meters indicate the distortion percentage directly on a meter scale and require no calculations.

It is often not enough to know which harmonics are present in a distorted alternating current or voltage and what their amplitudes are; the phase angles between the fundamental and individual harmonics also must be known (for *phase*, see Sec. 1.8) In this connection, an exhaustive study of a distorted wave requires a *Fourier analysis*, which involves the use of higher mathematics and sophisticated modern instruments. For most practical purposes, however, distortion measurements made with a wave analyzer, simple distortion meter, or oscilloscope (employing the schedule method*) will suffice.

---

*Many electronic engineering handbooks and textbooks give detailed instructions for use of this method.

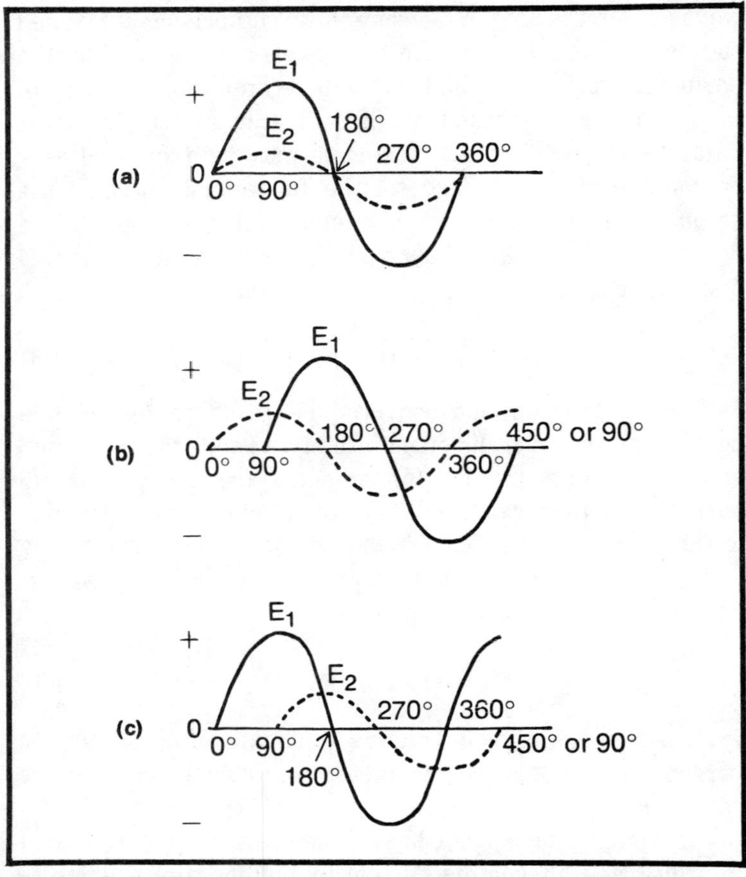

Fig. 1-4. Basic phase relationships of voltages and currents: (a) in phase, (b) leading phase, (c) lagging phase.

It can be shown mathematically, and also by the practical mixing of signals, that any nonsinusoidal wave is the combination of a certain number of sine waves of various frequencies (harmonics) and amplitudes. Thus, a square wave is the combination of a fundamental sine-wave frequency and numerous odd-numbered harmonics, a sawtooth wave is the combination of a fundamental sine-wave frequency and numerous even- and odd-numbered harmonics, etc. The more harmonics present, the more closely the complex wave approximates its ideal shape. The frequency of the complex wave itself is the same as the fundamental frequency.

1.8 PHASE

The alternations of two separate currents or voltages fall into one of three categories: they may be in step with each other; those of one may be ahead of those of the other; or those of one may be behind those of the other. This condition of being in or out of step is termed *phase relationship*. The three situations just cited—*in phase*, *leading phase*, and *lagging phase*—are illustrated in Fig. 1-4, which shows the relationship of two voltages that are in phase and out of phase. These figures serve to illustrate the general conditions; there are, of course, almost limitless combinations of out-of-phase quantities.

In Fig. 1-4(a), voltages E_1 and E_2 reach all of their values at the same instants and so are in phase. Their *phase difference* thus is zero degrees. In Fig. 1-4(b), E_2 reaches each of its values 90 degrees before E_1 does. In this case, E_2 is said to lead E_1, and their phase difference is 90 degrees. In Fig. 1-4(c), E_2 reaches each of its values 90 degrees after E_1 does. In this case, E_2 is said to lag E_1, and again their phase difference is 90 degrees. While a phase difference of 90 degrees is shown in Fig. 1-4(b) and (c), the angle can be anywhere between less than one degree to 360 degrees. (At exactly 360 degrees, of course, the in-phase condition of Fig. 1-4(a) is reestablished.) Here, we have followed the common practice of indicating phase in degrees, but it can be expressed also in radians and in seconds (time).

While two voltages are shown in each example in Fig. 1-4, phase relationships also exist between two currents, a voltage and a current, or a current and a voltage. Also, in Fig. 1-4, E_1 and E_2 are shown of different amplitude, but in practice the two components may be of the same amplitude or in opposite ratio to that shown here. It is important also to note that when harmonic frequencies are present in a wave, these components often are in different phase with each other *and* with the fundamental frequency.

The term *phase shift* refers to the change in phase relationship resulting from the flow of alternating current through certain devices or circuits. For example, at the input terminals of a certain "black box," current is in phase with

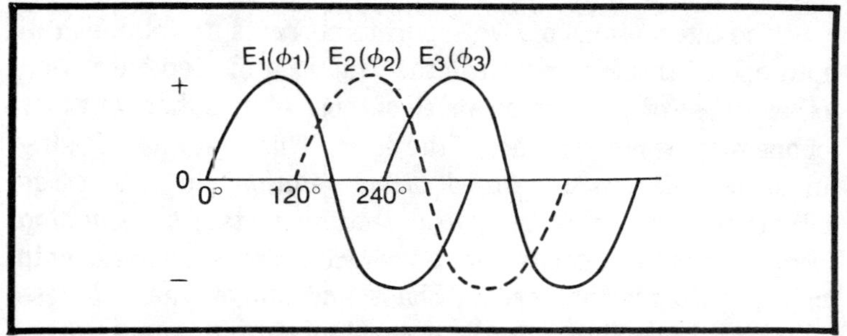

Fig. 1-5. Three-phase voltage. Three equal-amplitude voltages are spaced 120° apart.

voltage in an applied signal; but in the load connected to the output terminals, the current lags the voltage by 60 degrees. Thus, the black box has introduced a lagging phase shift. Current passing through a pure inductance *lags* applied voltage by 90 degrees, whereas current flowing into and out of a pure capacitance *leads* applied voltage by 90 degrees. In a common-cathode vacuum-tube circuit, common-emitter transistor stage, or common-source FET stage, the output signal voltage is 180 degrees out of phase with the input signal voltage. But in a cathode follower, emitter follower, or source follower, the output signal voltage is in phase with the input signal voltage.

Today, most AC energy is transmitted efficiently via three-phase systems, although much of it is converted to single-phase by service transformers located near the point of use. Where actual three-phase energy is available for use in electronic systems, it is valued for its uniform (nonpulsating) power, increased efficiency over single-phase energy in the operation of electrical machinery such as motors, and the ease with which it is filtered. The output of a three-phase generator consists of three equal-amplitude voltages spaced 120 degrees apart (see Fig. 1-5); thus, voltage E_1 starts at 0 degrees, E_2 at 120 degrees, and E_3 at 240 degrees). It is conventional to speak of each voltage as a phase (symbolized ϕ). In a balanced three-phase system, the total power is equal to 3 times the power ($EI \cos \theta$) in either one of the phases, which because of the phase differences is equal to:

$$P_T = 1.732\, E\, I \cos \theta \quad (1\text{-}8)$$

1.9 VECTOR REPRESENTATION OF AC COMPONENTS

It is often convenient to think of an alternating current or voltage in terms of a rotating vector. This concept is illustrated by the diagram in Fig. 1-6.

Here, the length of vector OA is equal or proportional to the maximum voltage or current value, E_{MAX} or I_{MAX}. This vector rotates counterclockwise from 0 to 360 degrees at the rate of $2\pi f$ radians per second. The vertical distance (AB) from the head of the vector to the horizontal axis is equal or proportional to the instantaneous voltage or current. As the vector rotates, AB increases positively from zero at 0 degrees to positive maximum at 90 degrees; then, as the vector rotates from 90 degrees to 180 degrees, AB decreases positively, returning to zero at 180 degrees. As the vector rotates from 180 degrees to 270 degrees, AB increases negatively from zero at 180 degrees to negative maximum at 270 degrees; then, as the vector rotates from 270 degrees to 360 degrees, AB decreases negatively from maximum at 270 degrees to zero at 360 degrees. One cycle thus has been completed and the events are ready to repeat themselves.

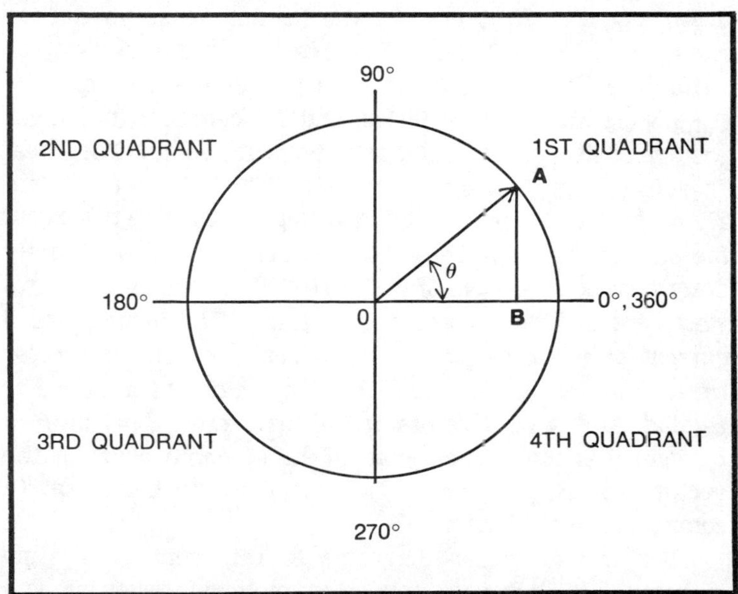

Fig. 1-6. Vector representation of AC components.

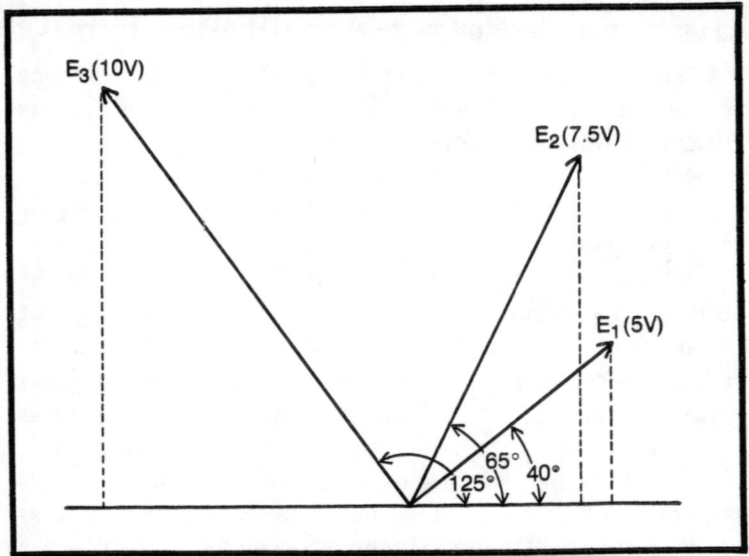

Fig. 1-7. Vector diagram of out-of-phase components.

The vector AB is proportional to the sine of the angle θ. Indeed, when the diagram is based on a unit circle, AB $=$ sin θ. It follows that in the latter case, 0B $=$ cos θ. Thus, when 0A is drawn equal to E_{MAX} or I_{MAX}, the instantaneous voltage or current AB $=$ 0Asin θ. This, of course, is just another way of writing: $e = E_{MAX} \sin \theta$, or $i = I_{MAX} \sin \theta$ (see Eq. 1-2). Component AB is zero at 0, 180, and 360 degrees; maximum positive at 90 degrees; and maximum negative at 270 degrees. Therefore: $\sin 0° = \sin 180° = \sin 360° = 0$; $\sin 90° \sin 270° = 1$. Thus, the periodically varying length of AB traces out the sine of the angle from 0 to 360 degrees and accurately describes the sine wave of Fig. 1-1(a). The following general statement describes these relationships: *The instantaneous current or voltage equals the product of a rotating vector times the sine of the angle through which the vector has rotated.* At any positive position of the vector, $E_{MAX} \sin \theta$ or $I_{MAX} \sin \theta$ is the vertical component (y component) of the vector, and $E_{MAX} \cos \theta$ or $I_{MAX} \cos \theta$ is the horizontal (**x** component) of the vector.

The use of vector diagrams to rep sent alternating currents and voltages is a convenient method for showing both magnitude and phase of these components. One could, of course, plot the waveforms to scale, but the vector diagram saves time and labor. Figure 1-7 is a vector diagram of three out-of-phase voltages. Here, E_1 is 5V at 40 degrees, E_2 7.5V at 65 degrees, and E_3 10V at 125 degrees. The vectors are drawn to scale to indicate the magnitude of these components. The same sort of diagram would be employed with three currents.

Each of these voltage vectors has a horizontal (**x**) component and a vertical (**y**) component. Also, there is a total **x** component ($E_{TOTAL\ X}$) and and total **y** component ($E_{TOTAL\ Y}$) which can be determined from the data presented by the diagram. Then, there is the single voltage (E_X) generated by the three out-of-phase components (E_1, E_2, and E_3) which is the resultant of $E_{TOTAL\ X}$ and $E_{TOTAL\ Y}$. Finally, there is the phase angle ϕ of E_X. The following schedule shows how these various voltages and the phase angle of E_X are calculated.

$$E_{1X} = 5 \cos 40° = 5(0.77604) = 3.88V$$
$$E_{2X} = 7.5 \cos 65° = 7.5(0.42262) = 3.17V$$
$$E_{3X} = 10 \cos 125° = 10(-0.57358) = -5.73V$$
$$E_{TOTAL\ X} = 3.88 + 3.17 - 5.73 = 1.32V$$
$$E_{1Y} = 5 \sin 40° = 5(0.64279) = 3.21V$$
$$E_{2Y} = 7.5 \sin 65° = 7.5(0.90631) = 6.79V$$
$$E_{3Y} = 10 \sin 125° = 10(0.81915) = 8.19V$$
$$E_{TOTAL\ Y} = 3.21 + 6.79 + 8.19 = 18.19V$$
$$\phi = \text{arc tan } 18.19/1.32 = \text{arc tan } 13.78 = 85.85°$$
$$E_X = E_{TOTAL\ Y}/\sin \theta = 18.19/\sin\ 85.85° = 18.19/0.99738 = 18.14V$$

1.10 AC IN RESISTANCE

A pure resistance (R) introduces no phase shift. Consequently, when as AC voltage is applied to a pure resistance, the resulting current flow through the resistance is in phase with the voltage. Figure 1-8(a) illustrates this action. Similarly, when an alternating current flows through a resistance, the resulting voltage drop across the resistance is in phase with the current.

Pure resistance consumes power but is not frequency-dependent in its action. There is nothing in a pure resistance that causes it to change, with frequency, the amount of opposition it offers to current flow. This is not true

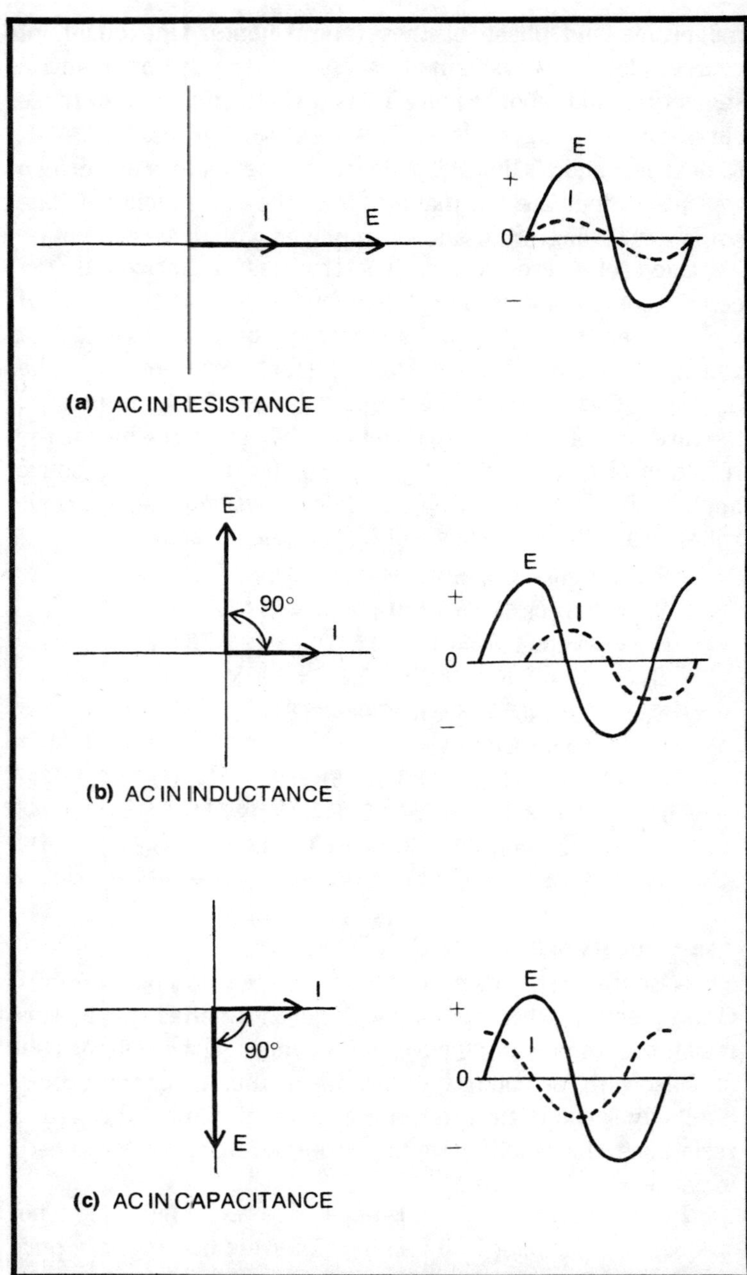

Fig. 1-8. Current/voltage/phase relationships with voltage applied to (a) a pure resistance, (b) a pure inductance, (c) a pure capacitance.

of *reactance* (X), which is a frequency-dependent opposition to current flow. Unlike resistance, pure reactance consumes no power. The kinds of reactance are described in Sec. 1.11, 1.12, and 1.13.

In a pure resistance, current is directly proportional to voltage and is inversely proportional to resistance, as shown by Ohm's law:

$$I = E/R, E = IR, R = E/I \tag{1-9}$$

where I is in amperes, E in volts, and R in ohms.
Although Ohm's law in this form is commonly associated with DC, it applies to AC as well, so long as the resistance is considered pure. (Ohm's law for AC circuits is often written with Z replacing the R; thus: $I = E/Z, E = IZ$, and $Z = E/I$.)

1.11 AC IN INDUCTIVE REACTANCE

When a voltage is applied to a pure inductance (L), current cannot flow immediately because it is opposed by a voltage of opposite polarity—the counter emf generated by the moving magnetic field of the inductor. The current reaches its maximum value some time after the voltage has been applied. Voltage applied to an inductance therefore leads current, as shown in Fig. 1-8(b), and it leads by 90 degrees in a *pure* inductance. (If unavoidable resistance is present, the phase angle is proportionately less than 90 degrees. The opposition thus offered by an inductance is termed *inductive reactance* (X_L).

For a given value of inductance, the strength of the counter emf is proportional to the rate of change of the applied voltage. Therefore, the higher the frequency, the higher the counter emf and the higher the reactance. The effective value of the induced counter emf is $E = 2\pi fLI$. Therefore, the formula for inductive reactance is:

$$X_L = \omega L = 2\pi fL \tag{1-10}$$

where X_L is in ohms, f in hertz, and L in henrys.

Example 1-6. A 15-henry (15H) inductor is operated in a 400 Hz circuit. Neglecting any inherent resistance, calculate the reactance at that frequency.

From Eq. 1-10,

$$X_L = 2\pi(400)15$$
$$= 37{,}699\Omega$$
$$= 3.7699K$$

A pure inductance consumes no power, since power stored in the expanding magnetic field during one quarter-cycle of AC is returned to the circuit by the collapsing magnetic field during the following quarter-cycle. In a pure inductive reactance, current is directly proportional to voltage and inversely proportional to reactance, as shown by Ohm's law:

$$I = E/X_L, E = IX_L, X_L = E/I \qquad (1\text{-}11)$$

where I is in amperes, E in volts, and X_L in ohms.
The sign of inductive reactance is positive.

Example 1-7. A 60 Hz sinusoidal current of 10 mA rms flows through a 2.5 mH inductor. Assuming that this is a pure inductance, calculate the voltage drop in millivolts across the inductor.

Here, 10 mA = 0.01A, and 2.5 mH = 0.0025H. From Eq. 1-10,

$$X_L = 2(3.1416)60(0.0025)$$
$$= 0.942\Omega$$

From Eq. 1-11, And $E = IX_L$

$$= 0.01(0.942)$$
$$= 0.00942V$$
$$= 9.42 \text{ mV}$$

1.12 AC IN CAPACITIVE REACTANCE

When a voltage is applied to a pure capacitance (C), as to an ideal lossless capacitor, a current flows into the capacitor, decreasing in value until the capacitor becomes fully charged, whereupon the flow stops. The voltage across the capacitor thus is zero when the current is maximum, and vice versa. Current flowing into a capacitor is proportional to the rate of change of voltage; for an AC voltage, this rate of change is maximum when the cycle is passing through zero, and is zero when the cycle is maximum. Voltage across a pure capacitance therefore lags current. From the other point of view, current leads voltage—see Fig. 1-8(c). The current leads by 90 degrees. If unavoidable resistance is present, the phase angle is proportionately less than 90 degrees. The opposition thus offered by a capacitance is termed *capacitive reactance* (X_C). For a given capacitance and voltage, the higher the frequency, the lower the reactance. The effective value of capacitor current $I = 2\pi fCE$. Therefore, the formula for capacitive reactance is:

$$X_C = 1/\omega C = 1/(2\pi fC) \tag{1-12}$$

where X_C is in ohms, f in hertz, and C in farads.

Example 1-8. A 0.0025 μF capacitor is operated in a 1 MHz circuit. Calculate its reactance in ohms at that frequency.

Here, 0.0025 $\mu\text{F} = 2.5 \times 10^{-9}\text{F}$ and 1 MHz $= 10^6$ Hz. From Eq. 1-12,

$$\begin{aligned} X_C &= \frac{1}{2 \times 3.1416 \times 10^6 \times (2.5 \times 10^{-9})} \\ &= 1/0.01571 \\ &= 63.7\Omega \end{aligned}$$

A pure capacitance consumes no power, since power stored in the electrostatic field of the capacitor during one quarter-cycle, when the capacitor is charging, is returned to the circuit during the following quarter-cycle, when the capacitor is discharging. Alternating current flows in and out of a capacitor, not *through* it. In a pure capacitve reactance, current is directly proportional to voltage and inversely proportional to reactance, as shown by Ohm's law:

$$I = E/X_C, E = IX_C, X_C = E/I \tag{1-13}$$

where I is in amperes, E in volts, and X_C in ohms.
The sign of capacitive reactance , incidentally, is negative.

Example 1-9. A sinusoidal 1000 Hz signal of $5V_{RMS}$ is applied to a 50 pF* capacitor. Neglecting any inherent resistance, calculate the current in milliamperes that flows in and out of this capacitor.

Here, 50 pF $= 5 \times 10^{-11}$F; and from Eq. 1-12:

$$X_C = \frac{1}{2(3.1416)\,1000\,(5 \times 10^{-11})}$$

$$= 1/3.1416 \times 10^{-7}$$
$$= 3{,}183{,}091\Omega$$

1.13 COMBINED REACTANCE

Both kinds of reactance—inductive and capacitive—are often found in a single circuit. The opposition offered to the flow of alternating current is then the combined effect of the two reactances. When the two reactances are in series, the combined reactance is the algebraic sum of the two:

$$X = X_L - X_C \qquad (1\text{-}14)$$

where X, X_L, and X_C are all in the same units (ohms, kilohms, etc.)

But when the two reactances are in parallel,

$$X = (-X_L X_C)/(X_L - X_C) \qquad (1\text{-}15)$$

The dominant reactive component determines the nature of the combined reactance. Thus, where $X_L = 100\Omega$ and $X_C = 10\Omega$, $X = 100 - 10 = 90\Omega$ inductive. Similarly, where $X_L = 25\Omega$ and $X_C = 60\Omega$, $X = 25 - 60 = -35\Omega$ capacitive. At one frequency—termed the *resonant frequency* (f_R)—the inductive reactance equals the capacitive reactance and, because of the difference in sign, the two cancel each other, leaving no reactance in the circuit. In that case, $\omega L = 1/\omega C$; and, when the values of L and C are known, the equivalent equation $2\pi fL = 1/2\pi fC$ can be rewritten to solve for f, the

---

*The abbreviation *pF* stands for *picofarads*, which is the equivalent of 10^{-12}F or $10^{-6}\mu$F.

resonant frequency: $$f = \frac{1}{2\pi\sqrt{LC}} \quad (1\text{-}16)$$

where f is in hertz, L in henrys, and C in farads.

The inductor and capacitor are connected in series in a *series-resonant circuit*; they are connected in parallel in a *parallel-resonant circuit*.

Example 1-10. Calculate the resonant frequency in kilohertz of 350 pF and 175 μH in combination.

Here, 350 pF = 3.5×10^{-10}F, and 175 μH = 1.75×10^{-4}H. From Eq. 1-16:

$$f = 1/2 \times 3.1416\sqrt{3.5 \times 10^{-10}(1.75 \times 10^{-4})}$$

$$= 1/6.2832\sqrt{6.12 \times 10^{-14}}$$

$$= \frac{1}{6.2832 \times (2.475 \times 10^{-7})}$$

$$= \frac{1}{1.55 \times 10^{-6}}$$

$$= 645{,}161 \text{ Hz}$$
$$= 645.16 \text{ kHz}$$

From a rewritten form of Eq. 1-16, the capacitance required to resonate a given inductance at a selected frequency is:

frequency is: $$C = \frac{1}{4\pi^2 f^2 L} \quad (1\text{-}17)$$

where C is in farads, f in hertz, and L in henrys.

Example 1-11. What value of capacitance in microfarads is required to resonate a 10H inductor at 500 Hz?

From Eq. 1-17:

$$C = \frac{1}{4 \times 3.1416^2 \times 500^2 \times 10}$$

$$= \frac{}{39.48 \times 250{,}000 \times 10}$$

$$1/98{,}700{,}000$$
$$1.01 \times 10^{-8}\,\mathrm{F} =$$
$$0.0101\ \mu\mathrm{F}$$

Similarly, with the aid of another rewritten form of Eq. 1-16, the inductance required to resonate a given capacitance at a selected frequency is:

$$L = \frac{1}{4\pi^2 f^2 C} \qquad (1\text{-}18)$$

where L is in henrys, f in hertz, and C in farads.

Example 1-12. What value of inductance in millihenrys is required to resonate a 10 pF capacitor at 3500 kHz?

Here, 10 pF $= 10^{-11}$F and 3500 kHz $= 3.5 \times 10^6$ Hz. From Eq. 1-18:

$$L = \frac{1}{4 \times 3.1416^2 \times (3.5 \times 10^6)^2 \times 10^{-11}}$$

$$= \frac{1}{39.48(1.225 \times 10^{13})10^{-11}}$$

$$= \frac{1}{39.48(1.225 \times 10^2)}$$

$$= \frac{1}{4.8363 \times 10^3)}$$

$$= 2.07 \times 10^{-4}\mathrm{H} \text{ --}$$
$$= 0.207\ \mathrm{mH}$$

It is important to remember that a given capacitance or inductance offers a different amount of reactance to the fundamental frequency and to each of the harmonics in a complex wave. For example, at the second harmonic, capacitive reactance is half the value at the fundamental frequency, and inductive reactance is twice the value at the fundamental frequency; at the third harmonic, capacitive reactance is one-third, and inductive reactance is three times; etc. Consequently, when complex voltage waveform is applied to a reactance, the resulting current can have a quite different waveshape because of the different amounts of attenuation of the component frequencies.

1.14 AC COMBINED WITH DC

Frequently, an alternating current is mixed with a steady direct current, or an alternating voltage is mixed with a steady direct voltage. This situation is found in the input and output circuits of tube and transistor amplifiers (where the DC is a bias current or voltage, and the AC is the signal riding on the bias) and in the unfiltered output of rectifiers (where the AC is the ripple).

Figure 1-9 shows two examples. In the upper trace, an AC voltage alternates about +1V as a mean, rising to +1.5V on

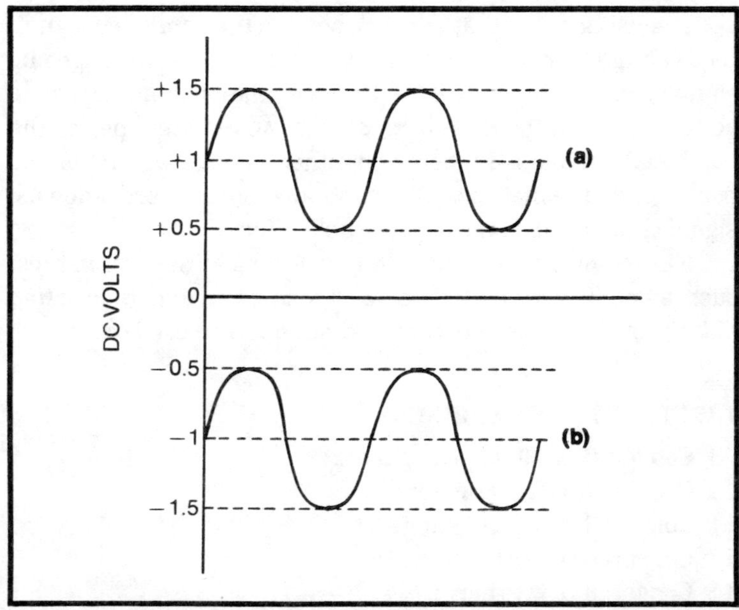

Fig. 1-9. Two examples of AC superimposed on DC. In (a) the alternating voltage is superimposed on +1V dc; in (b) the alternating voltage is superimposed on −1V dc.

positive peaks and falling to +0.5V on negative peaks. In the lower trace, an AC voltage of the same intensity alternates about −1V as a mean, falling to −1.5V on negative peaks and rising to −0.5V on positive peaks. In each instance, the wave is composed of a series of instantaneous DC values obtained by fluctuating the DC in some way (in a vacuum-tube amplifier, for example, an AC grid voltage swings the DC plate current up and down to produce the AC-on-DC signal).

Regardless of the instantaneous or average values of DC involved, the AC component exhibits only the conventional AC values—voltage or current—indicated by its dimensions. The rms value of each of the two waves in Fig. 1-9, for example, is $0.707(0.5) = 0.353$V, and it makes no difference whether the mean value is +1V, as in the upper figure, or −1V, as in the lower figure. Therefore, when the AC component is extracted from the mixture, as through capacitor coupling or transformer coupling, only this AC component, and none of the DC, is available in the output. The AC may be sinusoidal or nonsinusoidal.

It must be noted that at every point in the *combined* signal, the voltage (or current) is the sum of the average DC component (here, +1V or −1V) and the instantaneous AC voltage at that point. Thus, at the AC voltage peak, the combined voltage is higher than either the average DC or the peak AC, and sometimes this can cause circuit breakdowns, signal clipping, and other undesirable effects.

This combination of AC and DC goes under several names, such as *composite voltage* or *composite current*, *fluctuating voltage* or *fluctuating current*, and AC *superimposed on* DC.

1.15 PRACTICE EXERCISES

1.1 Convert 250,500 Hz to megahertz.

1.2 Convert 10 GHz to megahertz.

1.3 Convert 3.55 MHz to kilohertz.

1.4 Convert 60 Hz to kilohertz.

1.5 Convert 8 GHz to hertz.

1.6 Calculate the period in microseconds of a 5000 kHz standard-frequency signal.

1.7 Calculate the period in milliseconds of the 60 Hz power-line frequency.

1.8 Calculate the period in seconds of the 1540 kHz standard broadcast frequency.

1.9 Calculate the period in seconds of the 50 Hz power-line frequency.

1.10. Calculate the period in microseconds of a 1000 Hz audio test frequency.

1.11. Calculate the period in milliseconds of the 4000 kHz amateur frequency.

1.12. Calculate the period in microseconds of the 540 kHz standard broadcast frequency.

1.13. Calculate the period in seconds of the 27.125 MHz (channel 14) Citizens Band frequency.

1.14. Calculate the period in milliseconds of the 10.7 MHz FM intermediate frequency.

1.15. Calculate the period in microseconds of the 57 MHz center frequency of TV channel 2.

1.16. Calculate the period in seconds of a 1 GHz microwave signal.

1.17. Calculate the period in milliseconds of a 0.3 GHz microwave signal.

1.18. Calculate the period in microseconds of an 8 GHz microwave signal.

1.19. Calculate the frequency in hertz corresponding to a period of 0.01s.

1.20. Calculate the frequency in kilohertz corresponding to a period of 0.00015s.

1.21. Calculate the frequency in megahertz corresponding to a period of 10^{-5}s.

1.22. Calculate the frequency in gigahertz corresponding to a period of 10^{-10}s.

1.23. Calculate the frequency in hertz corresponding to a period of 8.33 ms.

1.24. Calculate the frequency in kilohertz corresponding to a period of 0.5 ms.

1.25. Calculate the frequency in megahertz corresponding to a period of 0.001 ms.

1.26. Calculate the frequency in gigahertz corresponding to a period of 2×10^{-3} ms.

1.27. Calculate the frequency in hertz corresponding to a period of 1000 μs.

1.28. Calculate the frequency in kilohertz corresponding to a period of 70 μs.

1.29. Calculate the frequency in megahertz corresponding to a period of 10 μs.

1.30. Calculate the frequency in gigahertz corresponding to a period of 0.005 μs.

1.31. A certain sine wave has a maximum value of 162.6V. Calculate the instantaneous voltage at 45 degrees.

1.32. A certain sine wave has a maximum value of 3V. Calculate the instantaneous voltage at 260 degrees.

1.33. A certain 1000 Hz sine wave has a maximum value of 10V. Calculate the instantaneous voltage at the 0.25 ms point.

1.34. A certain 60 Hz sine wave has a maximum value of 162.6V. Calculate the instantaneous voltage at the one second point.

1.35. A certain 1 MHz sine wave has a maximum value of 1V. At which instants in microseconds in the first cycle will the instantaneous voltage be −0.707V?

1.36. A certain 400 Hz sine wave has a maximum value of 8.91V. At what instant in milliseconds in the first cycle will the instantaneous voltage be +4.455V?

1.37. A certain sine wave has a maximum value of 10V. At the 15 μs point, the instantaneous voltage is 9.09V. Calculate the frequency in hertz of this wave.

1.38. A certain 1250 Hz sine wave has an instantaneous voltage of −5V at the 0.5 ms point in the cycle. Calculate the maximum voltage of this cycle.

1.39. In a 2.5 MHz sine-wave cycle, at which points in microseconds do the following voltages occur: (a) positive maximum; (b) negative maximum?

1.40. A certain sine-wave cycle has maximum positive voltage at the 1.25 ms point. Calculate the frequency of this wave.

1.41. Convert 39.5 degrees to radians.

1.42. Convert 5 degrees 15 minutes to radians.

1.43. Convert 5.4 radians to degrees.

1.44. Convert 1.047 radians to degrees.

1.45. What is the angle in radians at the 10 μs point in a 12.5 kHz sine-wave cycle?

1.46. What is the angle in radians at the 1.67 ms point in a 60 Hz sine-wave cycle?

1.47. At any frequency, what is the angle in radians in the sine-wave cycle at (a) maximum positive voltage; (b) maximum negative voltage?

1.48. For a 1000 Hz sine-wave cycle, express the angle in degrees when $t = 0.5$ ms.

1.49. For a 10 MHz sine-wave cycle, express the angle in degrees when $t = 0.075\ \mu s$.

1.50. Calculate the angular velocity (ω) for the following often used frequencies: (a) 40 Hz, (b) 125 Hz, (c) 800 Hz, (d) 100 kHz, (e) 540 kHz, (f) 1380 kHz, (g) 1.875 MHz, (h) 10.7 MHz, (i) 27.085 MHz, (j) 54 MHz.

1.51. What frequency in kilohertz is required for a desired angular velocity of 1000?

1.52. Calculate the rms value corresponding to a maximum voltage of 15V.

1.53. Calculate the rms value corresponding to a maximum voltage of 2.37 μV.

1.54. Calculate the average value corresponding to a maximum voltage of 6.9V.

1.55. Calculate the average value corresponding to a maximum voltage of 10 mV.

1.56. Calculate the rms value corresponding to an average voltage of 3.3V.

1.57. Calculate the rms value corresponding to an average voltage of 0.00015V.

1.58. Calculate the maximum value corresponding to an rms voltage of 50V.

1.59. Calculate the maximum value corresponding to an rms voltage of 1 μV.

1.60. Calculate the average value corresponding to an rms voltage of 510V.

1.61. Calculate the average value corresponding to an rms voltage of 38 mV.

1.62. In the test of a certain oscillator performed with a wave analyzer, the following signal voltages are observed: fundamental, 1V; second harmonic, 1 mV; third harmonic, 0.25 mV; and fourth harmonic, 0.1 mV. Calculate the harmonic strength in percent for (a) 2nd harmonic; (b) 3rd harmonic; (c) 4th harmonic.

1.63. From the data in exercise 1.62, calculate the total distortion in percent.

1.64. In the test of a certain amplifier performed with a distortion meter, the combined voltage is 2.2V and the total harmonic voltage is 1.45 mV. Calculate the total distortion in percent.

1.65. An audio generator is being adjusted for an acceptable total distortion of 0.25%. If the output of the generator is set to 1V, what must be the output voltage in millivolts of the distortion-measuring circuit for this percentage?

1.66. Calculate the counter emf in volts generated in a 30H inductor carrying 100 mA at 120 Hz.

1.67. Calculate the counter emf in volts generated in a 2.5 mH inductor carrying 1 mA at 1 MHz.

1.68. Calculate the 120 Hz reactance of a 15H inductor.

1.69. Calculate the 1 MHz reactance of a 100 μH inductor.

1.70. What inductance is required for 20K reactance at 1000 Hz?

1.71. At what frequency in hertz will a 20H inductor have a reactance of 10K?

1.72. Calculate the voltage drop in volts across a 2H inductor carrying 125 mA at 400 Hz.

1.73. Calculate the current in microamperes passed by a 1 mH inductor when the applied voltage is 250 mV at 2 MHz.

1.74. What is the 1000 Hz reactance of an inductor that passes 0.5A for an applied voltage of 10V?

1.75. Calculate the inductance in henrys of the inductor in exercise 1.74.

1.76. Calculate the 1 MHz reactance in ohms of a 0.002 μF capacitor.

1.77. Calculate the 50 MHz reactance in ohms of a 25 pF capacitor.

1.78. Calculate the 120 Hz reactance in ohms of a 16 μF capacitor.

1.79. Calculate the 60 Hz reactance in megohms of a 25 pF capacitor.

1.80. At what frequency will a 2 μF capacitor have a reactance of 1000Ω?

1.81. Calculate the effective current in milliamperes through a 1 μF capacitor at 1000 Hz when the applied potential is 1V.

1.82. Calculate the voltage required to force a current of 3 mA through a 0.01 μF capacitor at 1000 Hz.

1.83. Calculate the 400 Hz reactance of a capacitor which passes 1 mA at 10V.

1.84. Calculate the capacitance in microfarads of the capacitor in exercise 1.83.

1.85. Calculate the voltage drop in millivolts across a 0.025 μF capacitor carrying 500 μA at 2000 kHz.

1.86. What capacitance in microfarads will be required to pass 0.25A relay current at 60 Hz when the applied voltage is 115V?

1.87. A 50Ω inductive reactance and a 10Ω capacitive reactance are connected in series. Calculate the combined reactance.

1.88. A 50Ω inductive reactance and a 10Ω capacitive reactance are connected in parallel. Calculate the combined reactance.

1.89. (a) Calculate the combined 120 Hz reactance offered by a 20H inductor and an 8 μF capacitor in series. (b) Is the combined reactance inductive or capacitive?

1.90. (a) Calculate the combined 1 MHz reactance offered by a 1 mH inductor and a 0.01 μF capacitor in parallel. (b) Is the combined reactance inductive or capacitive?

1.91. Calculate the resonant frequency in kilohertz of a circuit containing 0.02 μF and 2.5 mH.

1.92. Calculate the resonant frequency in megahertz of a circuit containing 365 pF and 100 μH.

1.93. What capacitance in microfarads is required to resonate a 5H inductor at 400 Hz?

1.94. What capacitance in picofarads is required to resonate a 2 μH inductor at 45 MHz?

1.95. What inductance in millihenrys is required to resonate a 0.05 μF capacitor to 3500 Hz?

1.96. What inductance in henrys is required to resonate a 0.25 μF capacitor to 180 Hz?

(Correct answers are to be found in Appendix D.)

2 Nature of Impedance

This chapter surveys impedance and examines its composition and various aspects of its nature. The subject matter extends that of Chapter 1 by progressing from the concept of reactance developed at the end of that chapter. Illustrative examples are offered to demonstrate the various methods of calculating impedance.

2.1 IMPEDANCE DEFINED

Impedance (Z) is the opposition offered to the flow of alternating current and is expressed in ohms (where applicable, the multiples and submultiples of the ohm also are used: *microhms, milliohms, kilohms, megohms, etc.*). In this respect, the behavior of impedance in an AC circuit is analogous to that of resistance in a DC circuit and is described by Ohm's law:

$$Z = E/I, \quad I = E/Z, \quad E = IZ \tag{2-1}$$

where Z is in ohms, E is in volts, and I is in amperes.

Example 2-1. When an emf of 10V RMS is applied to a certain two-terminal black box, a current of 0.75 mA flows. Calculate in kilohms the internal impedance of the black box.

Here, 0.75 mA = 0.00075A. From Eq. 2-1:

$$Z = E/I$$
$$= 10/0.00075$$
$$= 13{,}333 \text{ ohms}$$
$$= 13.33\text{K}$$

The similarity ends there, however, since impedance, unlike resistance, is frequency dependent and exhibits phase angle.

In a very broad sense, *impedance* denotes *any* opposition that is offered to AC. Such a definition would include resistance and reactance, which by themselves, of course, are not strictly impedances. It is for this reason that such terms as *resistive impedance* (for resistance) and *reactive impedance* (for reactance) are sometimes encountered.

2.2 COMPOSITION OF IMPEDANCE

Impedance (Z) is the combined effect of resistance (R) and reactance (X). The resistive component is 90 degrees out of phase with the reactive component, so R and X cannot simply be added arithmetically to give the impedance. The vector diagrams in Fig. 2-1 show how resistance and reactance combine to form impedance.

Figure 2-1(a) shows resistance and inductive reactance. Here, the impedance vector (Z) is the resultant—the vector sum—of the resistance vector (R) and the reactance vector (X_L). The phase angle of the resulting impedance is the angle θ between the impedance vector and the resistance vector.

Figure 2-1(b) shows resistance and capacitive reactance. Here, the X_C vector is drawn in the opposite direction of the X_L vector in Fig. 2-1(a) to show that the effect of capacitive reactance is opposite to that of inductive reactance. The impedance vector (Z) is the resultant—the vector sum—of the resistance vector (R) and the reactance vector (X_C). The phase angle of the resulting impedance is the angle θ between the impedance vector and the resistance vector.

In Fig. 2-1(c) there is combined reactance (X) consisting of inductive reactance (X_L) and capacitive reactance (X_C). This combined reactance $X = X_L - X_C$ (see Sec. 1.13, Ch. 1)

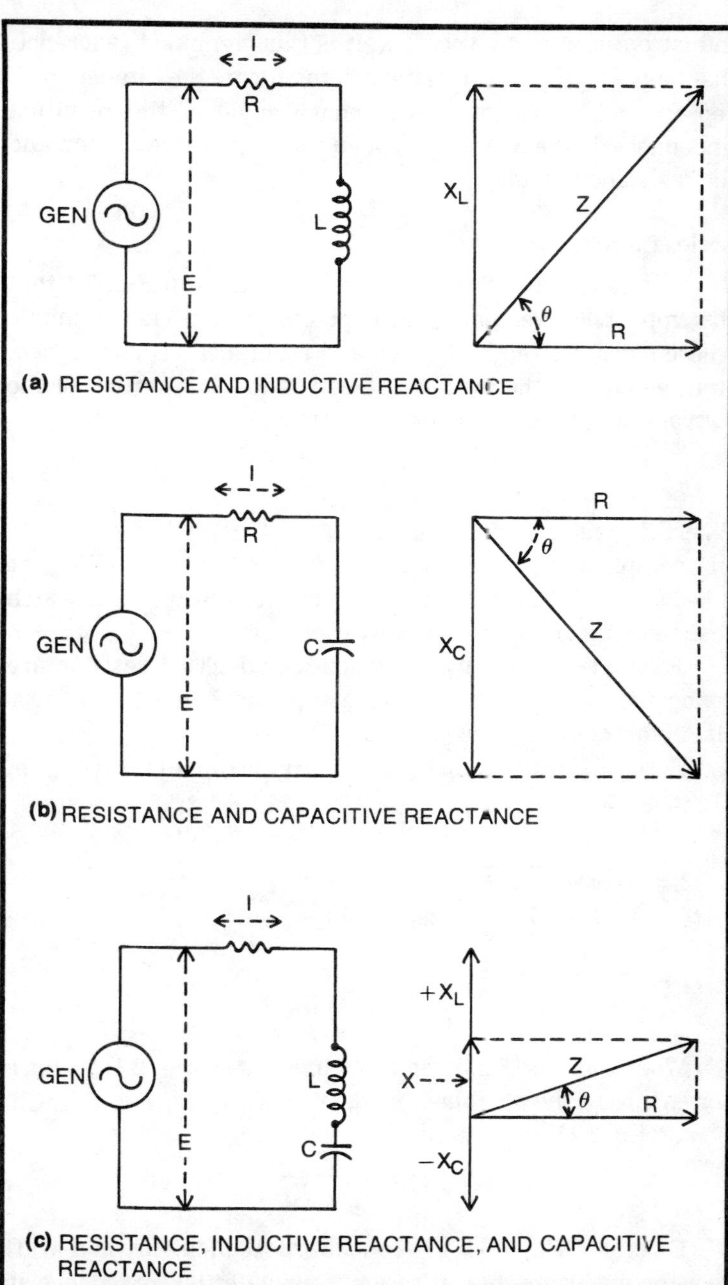

Fig. 2-1. Basic R-C-L-Z relationships with vectors showing how resistance and reactance combine to form impedance.

and is represented by vector **x**. It is this combined reactance that acts with the resistance to form the impedance, represented by vector **z**. The phase angle of the resulting impedance is the angle θ between the impedance vector and the resistance vector.

Series Circuits

It is easily seen from the three diagrams in Fig. 2-1 that the impedance vector is the hypotenuse of a right triangle whose sides are the resistance and reactance vectors. Since, from geometry, the hypotenuse equals the square root of the sum of the squares of the other two sides:

$$Z = \sqrt{R^2 + X^2} \tag{2-2}$$

Where Z, R, and X are in ohms.
In complex algebra, this is written $Z = R + jX_L$ or $Z = R - jX_C$. Equation 2-2 applies to circuits in which resistance and reactance are in series.

Example 2-2. A 0.1 μF capacitor and 1000Ω resistor are connected in series. Calculate the impedance in ohms (at 1000 Hz) of this combination.

Here, X_C for the 0.1 μF capacitor is 1591.5Ω (Eq. 1-12, Ch. 1). From Eq. 2-2:

$$\begin{aligned} Z &= \sqrt{1000^2 + 1591.5^2} \\ &= \sqrt{(1 \times 10^6) + (2.533 \times 10^6)} \\ &= \sqrt{3.533 \times 10^6} \\ &= 1879.6\Omega \end{aligned}$$

When there is combined reactance, as in Fig. 2-1(c), for a series circuit the combined value $X = X_L - X_C$ (Eq. 1-14, Ch. 1), and Eq. 2-2 is rewritten:

$$Z = \sqrt{R^2 + (X_L - X_C)^2} \tag{2-3}$$

Example 2-3. A 0.5 μF capacitor, 1H inductor, and 470Ω resistor are connected in series. Calculate the impedance in ohms (at 400 Hz) of this combination.

Here, X_L for the 1H inductor is 2513.2Ω (Eq. 1-10, Ch. 1) and X_C for the 0.5 μF capacitor is 795.8Ω (Eq. 1-12, Ch. 1). From Eq. 2-3:

$$\begin{aligned} Z &= \sqrt{470^2 + (2513.3 - 795.8)^2} \\ &= \sqrt{220{,}900 + 1717.5^2} \\ &= \sqrt{220{,}900 + 2{,}949{,}806} \\ &= \sqrt{3{,}170{,}706} \\ &= 1780.6\Omega \end{aligned}$$

Since the diagrams in Fig. 2-1 are right triangles, the solutions from trigonometry are easily applied. Thus, the tangent of the phase angle (θ) of the impedance, being equal to the opposite side divided by the adjacent side of the triangle, is equal to X/R:

$$\tan\theta = X/R = X_L/R = X_C/R \qquad (2\text{-}4)$$

where X, X_L, and X_C are in ohms.
When the reactance and resistance are known, the phase angle can be found:

$$\theta = \arctan X/R \qquad (2\text{-}5)$$

where θ is in degrees, and X and R are in ohms.
Likewise, $\sin\theta = X/Z$, and $\cos\theta = R/Z$. From this, $\theta = \arcsin X/Z = \arccos R/Z$.

Example 2-4. A 10 mH inductor and 56Ω resistor are connected in series. Calculate the phase angle of the impedance at 1000 Hz.

Here, X_L for the 10 mH inductor is 62.8Ω (Eq. 1-10, Ch. 1). From Eq. 2-5,

$$\begin{aligned} \theta &= \arctan 62.8/56 \\ &= \arctan 1.1214 \\ &= 48.275 \text{ degrees} \\ &= 48 \text{ degrees, } 16 \text{ minutes, } 30 \text{ seconds.} \end{aligned}$$

The total impedance of similar impedances connected in series is similar to the total resistance of resistors connected in series:

$$Z_T = Z_1 + Z_2 + Z_3 + \ldots Z_N$$

Parallel Circuits

When resistance and reactance are in parallel, the resulting impedance is:

$$Z = \frac{RX}{\sqrt{R^2 + X^2}} \tag{2-6}$$

where Z, R, and X are in ohms.

This formula is seen to resemble that for two resistances in parallel: $R_{EQ} = (R_1R_2)/(R_1 + R_2)$. But whereas in the resistance formula the product is divided by the sum, in the impedance formula (because of the difference between R and X) the product is divided by the *vector* sum.

Example 2-5. A 20H inductor and 5K resistor are connected in parallel. Calculate the impedance in kilohms (at 500 Hz) of this combination.

Here, X_L for the 20H inductor is 62,832Ω (Eq. 1-10, Ch. 1). From Eq. 2-6:

$$\begin{aligned} Z &= (5000 \times 62{,}832)/\sqrt{5000^2 + 62{,}832^2} \\ &= (3.142 \times 10^8)/\sqrt{(2.5 \times 10^7) + (3.95 \times 10^9)} \\ &= (3.142 \times 10^8)/\sqrt{3.975 \times 10^9} \\ &= \frac{3.142 \times 10^8}{6.305 \times 10^4} \\ &= 4893\Omega \\ &= 4.893\text{K} \end{aligned}$$

As with a parallel-resistance circuit with unequal resistances, the impedance of the parallel resistance/reactance circuit is less than either the resistance or the reactance.

For the parallel circuit, the phase angle of the impedance is:

$$\theta = \arc\tan R/X \tag{2-7}$$

where θ is in degrees, R in ohms, and X in ohms.

Note that this formula is the reciprocal of the one for the phase angle of the series circuit (Eq. 2-5).

Example 2-6. In the preceding example, $X_L = 62{,}832\Omega$ and $R = 5K$. Calculate the phase angle of the resulting 4893Ω impedance.

From Eq. 2-7,

$$\begin{aligned}\theta &= \text{arc tan } 5000/62{,}832 \\ &= \text{arc tan } 0.079577 \\ &= 4.549 \text{ degrees} \\ &= 4 \text{ degrees, } 32 \text{ minutes, } 56 \text{ seconds.}\end{aligned}$$

The equivalent impedance of similar impedances connected in parallel is similar to the equivalent resistance of resistors connected in parallel:

$$Z_{EQ} = \frac{1}{(1/Z_1 + 1/Z_2 + 1/Z_3 + \ldots + 1/Z_N)}$$

Full Depiction

Equations 2-1, 2-2, 2-3, and 2-6 give only the magnitude of impedance. In many applications, this quantity is all that is needed. A full expression of impedance, however, contains not only the magnitude, but also the phase angle (see Eq. 2-5 and 2-7 for the angle). For example: $Z/\theta = 35/26$ degrees 30 minutes denotes an impedance of 35Ω at a phase angle of 26 degrees and 30 minutes.

When the magnitude Z and phase θ of an impedance are given, the resistive (R) and reactive (X) components may be determined either graphically or through calculation. In the graphic solution (Fig. 2-2), the impedance vector **z** is drawn to scale forming the angle θ with the horizontal (resistance) axis. Then, projections are made from the tip of the **z** vector to the horizontal and vertical axes, as shown by the dotted lines. The resistance magnitude may then be measured along the horizontal axis, and the reactance magnitude along the vertical axis. The solution by calculation is based on simple right-triangle relationships from trigonometry:

$$R = Z \cos \theta, \text{ and} \tag{2-8}$$

$$X = Z \sin \theta \tag{2-9}$$

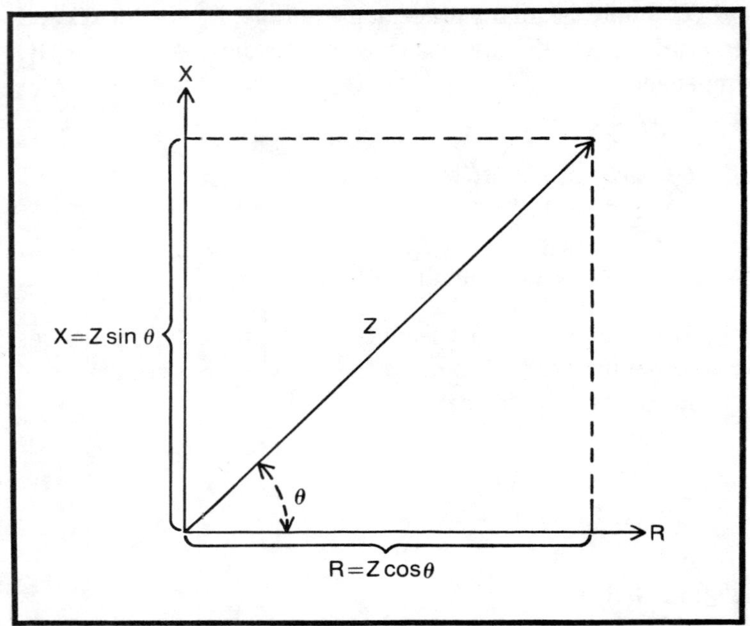

Fig. 2-2. Determination of resistance and reactance from impedance and phase angle. The resistance magnitude may be measured along the horizontal axis and the reactance along the vertical axis.

Example 2-7. A given impedance is 150Ω at 30 degrees. Calculate the resistive and reactive components.

Here, sin 30 degrees = 0.5 and cos 30 degrees = 0.866025. So, from Eqs. 2-8 and 2-9,

$$\begin{aligned} R &= 150(0.866025) \\ &= 129.9\Omega \\ X &= 150(0.5) \\ &= 75\Omega \end{aligned}$$

From Eqs. 2-8 and 2-9, it is apparent that impedance may be calculated in terms of resistance and phase angle *or* reactance and phase angle: $Z = R/\cos\theta$, and $Z = X/\sin\theta$.

2.3 UNIVERSALITY OF IMPEDANCE

Impedance is found everywhere in the world of electronics. This is because resistance and reactance tend to occur together, one often as a stray effect. Thus, a resistor can exhibit inherent capacitance and inductance, a capacitor can exhibit inherent resistance and inductance, and an inductor can exhibit inherent resistance and capacitance. It is stray resistance that causes losses in capacitors and inductors. In most well built components, the stray quantity is negligible when compared with the principal property. When the value of the stray is significant, however, the component or device must be handled as an impedance, not as a simple resistance or reactance.

Some of the familiar devices in which impedance is encountered are antennas and transmission lines; generators, motors, relays, and transformer windings; headphones, microphones, loudspeakers, and magnetic amplifiers; capacitors, inductors, saturable reactors, and resistors (inductively wound); tubes, transistors, semiconductor diodes, and rectifiers; and control devices.

2.4 IMPEDANCE OF COMMON BASIC CIRCUITS

Figure 2-3 shows eight common circuits with the formulas for their impedance and phase angle. These are basic arrangements in which resistance, capacitance, and inductance are assumed to be ideal. Several of these circuits invite special attention and are discussed individually.

Figure 2-3(e) shows an ideal *series-resonant circuit*. Depending upon the various values which inductance (L) and capacitance (C) may assume, the circuit may be resonant (exhibiting no reactance), nonresonant above the resonant frequency (exhibiting inductive reactance), or nonresonant below the resonant frequency (exhibiting capacitive reactance). The phase angle of the inductive reactance is +90 degrees, and that of the capacitive reactance −90 degrees; for frequencies off resonance, the angle is positive if X_C is larger than X_L, and is negative if X_L is larger than X_C. At resonance, since at this point $X_L = X_C$, angle θ is zero. The impedance at frequencies off resonance is equal to $X_L - X_C$ and is characterized by the dominant member of this expression. At resonance, therefore, Z is zero—which accounts for maximum current at resonance in series-resonant circuits.

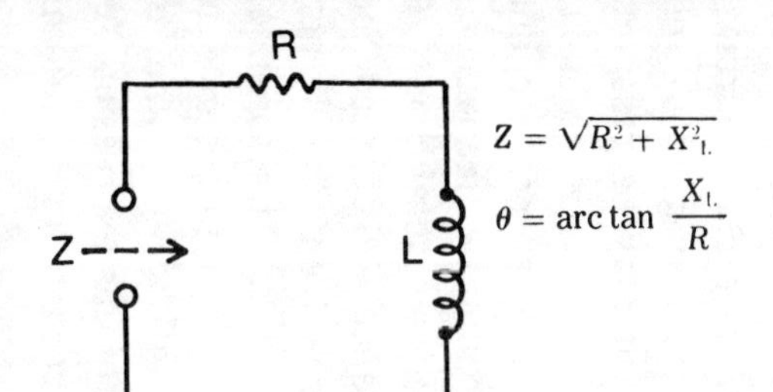

$$Z = \sqrt{R^2 + X_L^2}$$

$$\theta = \arctan \frac{X_L}{R}$$

(a) RESISTANCE AND INDUCTANCE IN SERIES

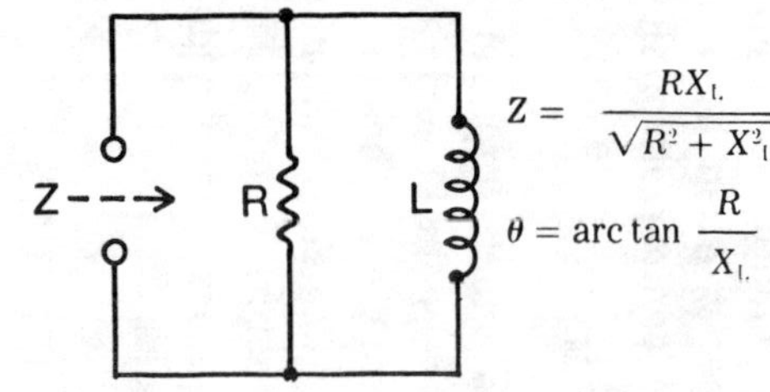

$$Z = \frac{RX_L}{\sqrt{R^2 + X_L^2}}$$

$$\theta = \arctan \frac{R}{X_L}$$

(b) RESISTANCE AND INDUCTANCE IN PARALLEL

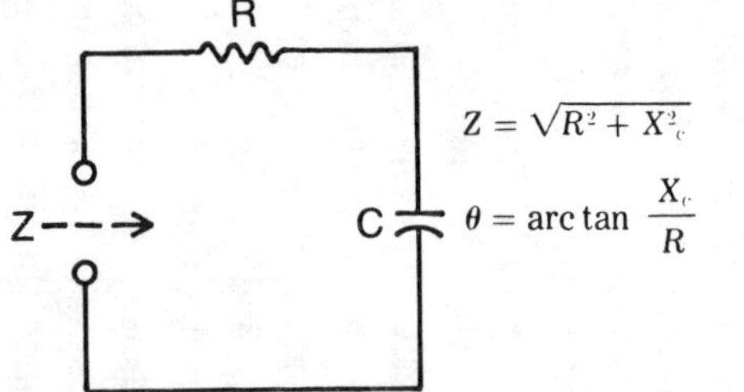

$$Z = \sqrt{R^2 + X_c^2}$$

$$\theta = \arctan \frac{X_c}{R}$$

(c) RESISTANCE AND CAPACITANCE IN SERIES

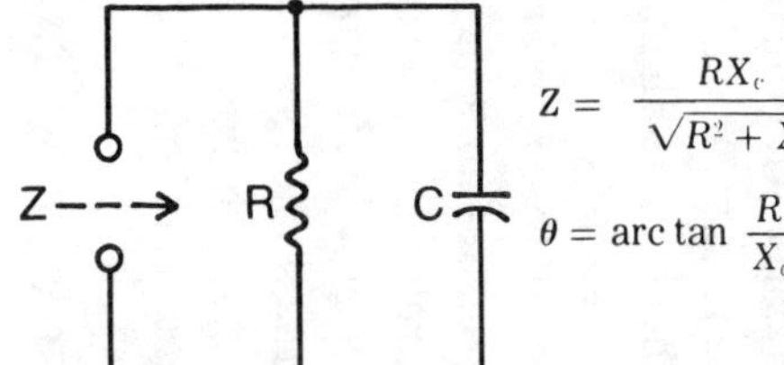

$$Z = \frac{RX_c}{\sqrt{R^2 + X^2}}$$

$$\theta = \arctan \frac{R}{X_c}$$

(d) RESISTANCE AND CAPACITANCE IN PARALLEL

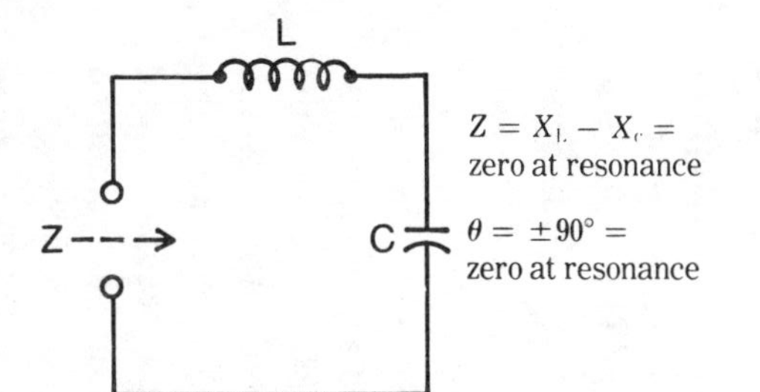

$Z = X_L - X_C =$
zero at resonance

$\theta = \pm 90° =$
zero at resonance

(e) INDUCTANCE AND CAPACITANCE IN SERIES

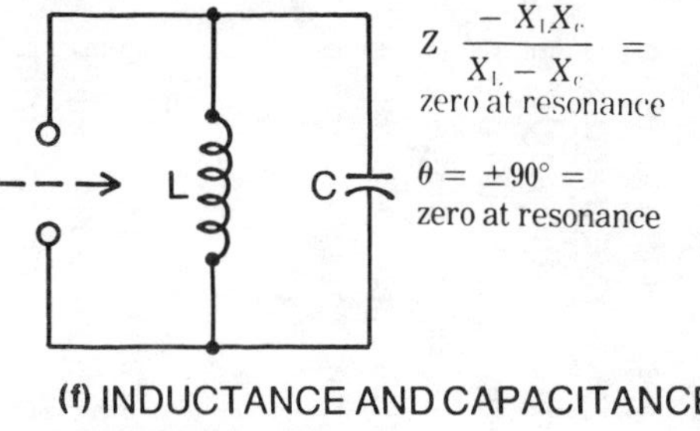

$Z \ \frac{-X_L X_C}{X_L - X_C} =$
zero at resonance

$\theta = \pm 90° =$
zero at resonance

(f) INDUCTANCE AND CAPACITANCE IN PARALLEL

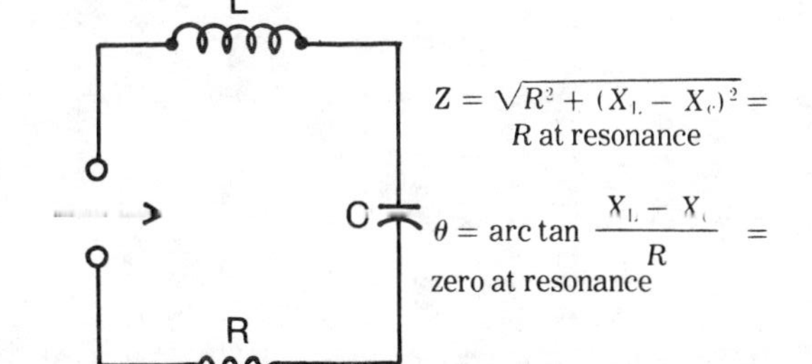

$Z = \sqrt{R^2 + (X_L - X_C)^2} =$
R at resonance

$\theta = \arctan \frac{X_L - X_C}{R} =$
zero at resonance

(g) INDUCTANCE, CAPACITANCE, AND RESISTANCE IN SERIES

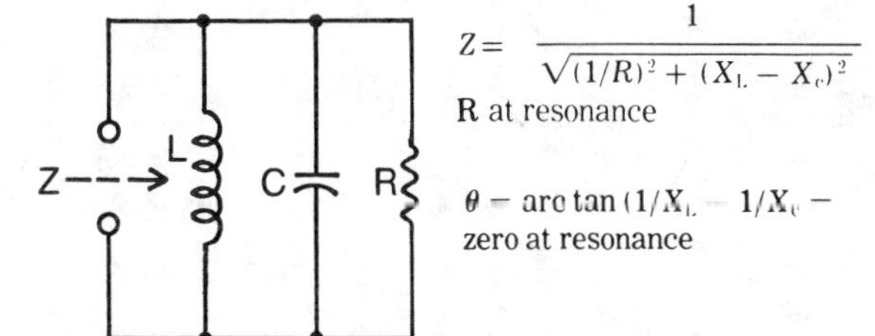

$Z = \frac{1}{\sqrt{(1/R)^2 + (X_L - X_C)^2}} =$
R at resonance

$\theta = \arctan (1/X_L - 1/X_C -$
zero at resonance

(h) INDUCTANCE, CAPACITANCE, AND RESISTANCE IN PARALLEL

Fig. 2-3. Eight common basic circuits with equations for determining impedance and phase angle.

Figure 2-3(f) shows an ideal *parallel-resonant* circuit. In this arrangement, unlike the series-resonant circuit described in the preceding paragraph, the impedance at resonance is infinite. This accounts for maximum voltage at resonance in parallel-resonant circuits. The phase angle of the inductive reactance is +90 degrees and that of the capacitive reactance is −90 degrees; for frequencies off resonance, the angle is positive if X_C is larger than X_L, and is negative if X_L is larger than X_C. At resonance, since at this point $X_L = X_C$, angle θ is zero. The impedance at frequencies off resonance is equal to $X_L - X_C$ and is characterized by the dominant member of this expression. At resonance, since here $X_L = X_C$, the denominator of the impedance formula in Fig. 2-3(f) is zero; therefore, the impedance is infinite.

While Figs. 2-3(e) and 2-3(f) show ideal series-resonant and parallel-resonant circuits, Figs. 2-3(g) and 2-3(h) show corresponding practical circuits. That is, each of the latter circuits contain resistance which occurs in practice in the form of losses in the inductor and capacitor. In the series-resonant circuit, Fig. 2-3(g), the off-resonance impedance is the vector sum of the resistance and combined reactance, and is capacitive below resonance and inductive above resonance. At resonance, the combined reactance is zero, and only the resistance is left in the circuit. Therefore, at resonance $Z = R$. Current in the practical series-resonant circuit is maximum at resonance, but is limited by resistance. The phase angle is determined by the ratio of the combined reactance to the resistance. This angle may have any value between zero degrees and nearly 90 degrees, depending upon the relative amounts of X_L, X_C, and R. At resonance, the phase angle is zero, since here $X_L = X_C =$ zero, and arc tan $0/R = 0$.

In the parallel-resonant circuit, Fig. 2-3(h), the off-resonance impedance is equal to the reciprocal of the vector sum of the reciprocal of the resistance and the combined reactance, and is inductive below resonance and capacitive above resonance. At resonance, the combined reactance ($X_L - X_C$) is zero and only the resistance (R) is left in the circuit. Therefore, at resonance, $Z = R$. The phase angle is determined by the relative amounts of X_L, X_C, and R,

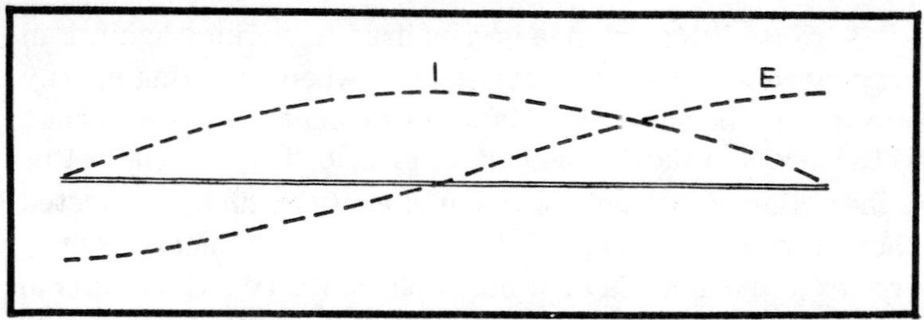

Fig. 2-4. Current and voltage distribution for a half-wave antenna operating at its fundamental frequency.

and may have any value between zero degrees and almost 90 degrees. At resonance this angle is zero, since here $X_L = X_C = 0$, and $\theta = \text{arc tan } R(1/X_L - 1/X_C) = \text{arc tan } R(0) = 0$.

2.5 IMPEDANCE OF LINEAR DEVICES

The impedance of devices which consist essentially of one or more straight wires, rods, or tubes (so-called *linear devices*) presents a special case. Prominent among such devices are *antennas* and RF *transmission lines*. In many instances the impedance of these devices is resistive.

Antennas

An operative antenna is characterized by a pattern of stationary standing waves along its length. This arrangement of loops and nodes constitutes a distribution of current I and voltage E along the length, as shown in Fig. 2-4 for a half-wave antenna operating at its fundamental frequency. By cutting or lengthening this figure, one can see what the resulting E and I distribution would be on antennas of different lengths, say quarter-wave and full-wave.

Note that current is maximum at the center of the wire, rod, or tube, and is zero at the ends, while voltage is zero at the center and maximum at the ends. At any point along the length of the antenna, the impedance Z is equal to the ratio of voltage to current (E/I) at that particular point. Thus, the impedance is very high at the ends (being theoretically infinite: $Z = E/I = E/0 = \infty$) and is very low at the center (being theoretically zero: $Z = E/I = 0/I = 0$).

A transmitting antenna is visualized as working against an impedance—the radiation resistance—when radiating energy into space. The value of radiation resistance (R_R) is governed by the height of the antenna above ground. The reason for this is the action of that part of radiated energy which is reflected back from the surface of the earth. This reflected energy arrives at the antenna in or out of phase with energy that is in the antenna. Depending upon how far the reflected energy has had to travel to reach the antenna, it either reduces or increases the apparent resistance because of this phase effect. Figure 2-5 shows a plot of theoretical values of radiation resistance at the center of a half-wave antenna in free space for various heights from zero to two wavelengths above

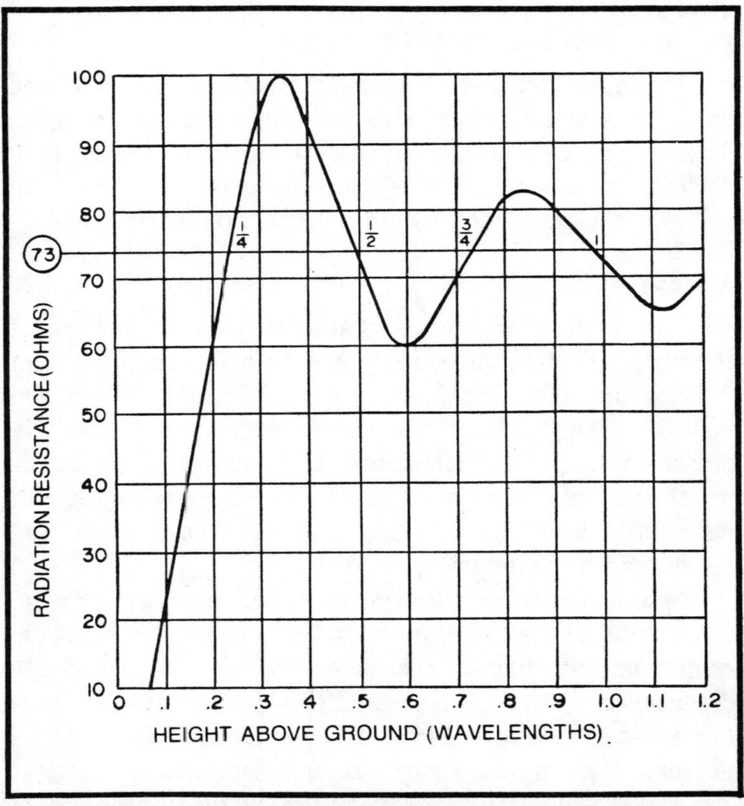

Fig. 2-5. Radiation resistance of a half-wave horizontal antenna plotted for various heights from zero to two wavelengths above ground.

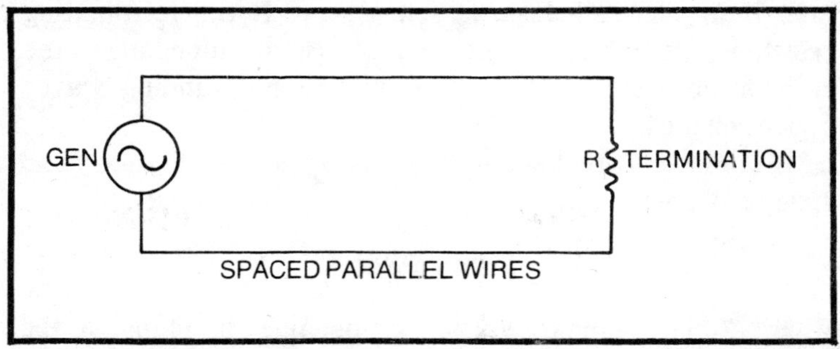

Fig. 2-6. Two-wire transmission line connected to an RF generator and a load resistor (termination).

ground. Observe that the higher the antenna, the more closely R_R approaches the theoretical value of 73.2Ω. At the ends of the antenna, R_R is several thousand ohms. In practical terms, the radiation resistance is that value of resistance which would, if it were inserted at the center of the antenna, dissipate energy equal to that ordinarily radiated from the antenna. And this is a legitimate concept, for radiated energy is, in effect, energy lost from the antenna.

Transmission Lines

The purpose of a *transmission line* is to conduct RF energy from one point (such as a generator) to another point (such as a load) with virtually no radiation from the line. In one of its simplest forms, this device consists of two parallel wires, with the spacing between the wires small compared with one wavelength. Figure 2-6 shows such a line connected to an RF generator at one end and to a load resistor (R) at the other end. Current flows in opposite directions in the two wires, so radiation from the line is effectively canceled. The line has distributed inductance and distributed capacitance, and from these properties the *characteristic impedance* (Z_0 or Z_C), neglecting the resistance of the wires, can be calculated:

$$Z_0 = \sqrt{L/C} \quad (2\text{-}10)$$

where Z_0 is in ohms, L in henrys, and C in farads. This quantity is termed *charteristic impedance*, since for a line of given dimensions it has the same E/I value at any point along the line. It is also called *surge impedance*. If the terminating resistance is equal to the characteristic impedance, the resistor absorbs all of the energy and no standing waves appear on the line.

For a two-wire line, Z_0 depends upon the diameter and spacing of the wires:

$$Z_0 = 276 \log_{10} 2S/d \qquad (2\text{-}11)$$

where Z_0 is the characteristic impedance in ohms, S the center-to-center spacing of wires in inches, d the diameter of wire in inches, and $\log_{10}$ the common logarithm.

Example 2-8. The diameter of No. 12 solid copper wire is 0.081 inch. Calculate the characteristic impedance of a two-wire line consisting of two No. 12 wires spaced six inches between centers.

From Eq. 2-11:

$$\begin{aligned} Z_0 &= 276 \log_{10}(2 \times 6)/0.081 \\ &= 276 \log_{10} 12/0.081 \\ &= 276 \log_{10} 148.15 \\ &= 276(2.1707) \\ &= 599.1\Omega \end{aligned}$$

Note: A pair of 12-gage wires with six-inch spacing is commonly called a 600Ω line.

A closer result (599.78Ω) is afforded by the equation $Z_0 = 120$ arc $\cosh[0.5(2S/d)]$, where Z_0, S, and d are in the same units as in Eq. 2-11 and *cosh* is the hyperbolic cosine.

The impedance of an insulated line is somewhat different from that of the open-air line just described. Thus, the three-eighth-inch wide "ribbon" used with TV antennas has an impedance of 300Ω.

Figure 2-7 shows the distribution of current and voltage on an unterminated quarter-wave line. From this distribution, it is evident that various impedances ($Z = E/I$) are available by tapping the line at appropriate points. This is an important convenience which will be considered later in Sec. 2-11, *Methods of Matching Impedance*.

Another well known transmission line is the *coaxial* type. This consists essentially of two concentric conductors, one

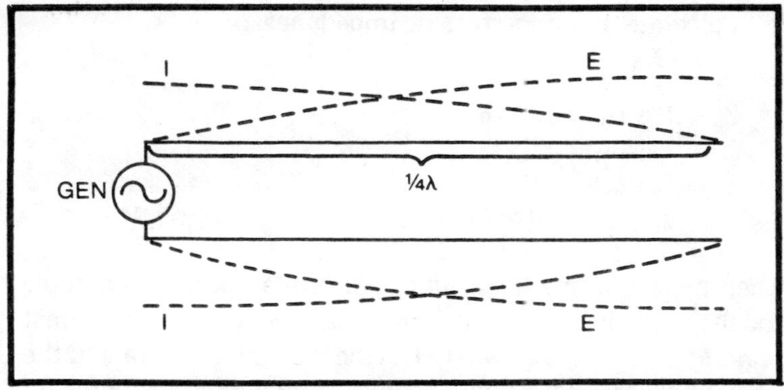

Fig. 2-7. Current and voltage distribution on an unterminated quarter-wave line.

being a central wire and the other a surrounding metal pipe (see Fig. 2-8). A coaxial line may be flexible or rigid. For an air-insulated coaxial line (inner conductor supported by spaced beads or washers), the characteristic impedance is:

$$Z_0 = 138 \log_{10} d_1/d_2 \qquad (2\text{-}12)$$

where Z_0 is the characteristic impedance in ohms, d_1 the inside diameter of the outer conductor in inches, d_2 the outside diameter of the inner conductor in inches, and $\log_{10}$ the common logarithm.

Example 2-9. The inner conductor of a certain air-insulated coaxial line is No. 12 copper wire whose outside diameter (OD) is 0.081 inch, and the inner diameter (ID) of the outer conductor is 0.25 inch.

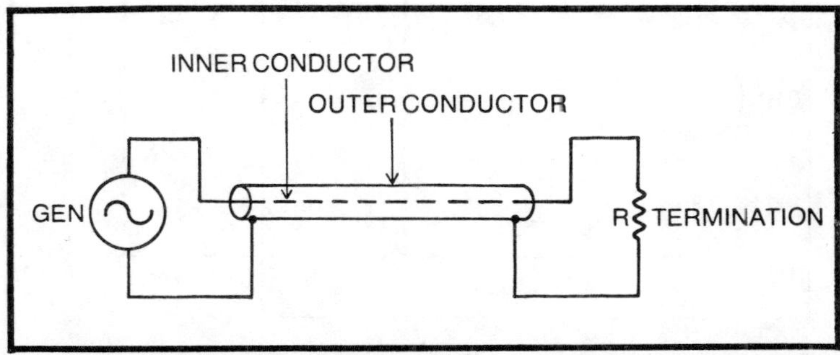

Fig. 2-8. Coaxial-type transmission line consisting of two concentric conductors connected between a generator and a load resistor.

Calculate the characteristic impedance.
From Eq. 2-12:

$$\begin{aligned} Z_0 &= 138 \log_{10} 0.25/0.081 \\ &= 138 \log_{10} 3.0964 \\ &= 138(0.489452) \\ &= 67.5\Omega \end{aligned}$$

When a coaxial line has continuous insulation between outer and inner conductors, the Z_0 value obtained with Eq. 2-12 must be multiplied by $1/\sqrt{k}$, where k is the dielectric constant of the insulating material. Polyethylene, a common insulator in coaxial lines, has a dielectric constant of 2.3 and requires a multiplier of $1/\sqrt{2.3} = 1/1.516 = 0.659$. Common impedances for commercial polyethylene-insulated coaxial cable are 50Ω, 52Ω, 53.5Ω, 73Ω, and 75Ω.

2.6 IMPEDANCE OF GENERATORS

All AC generators have impedance (Z_G). This impedance, however small, is often resistive and is considered to be in series with the generator (see Fig. 2-9). Because of the internal impedance, the terminal voltage (E_{TERM}) when the generator is delivering current to a load will be lower than the

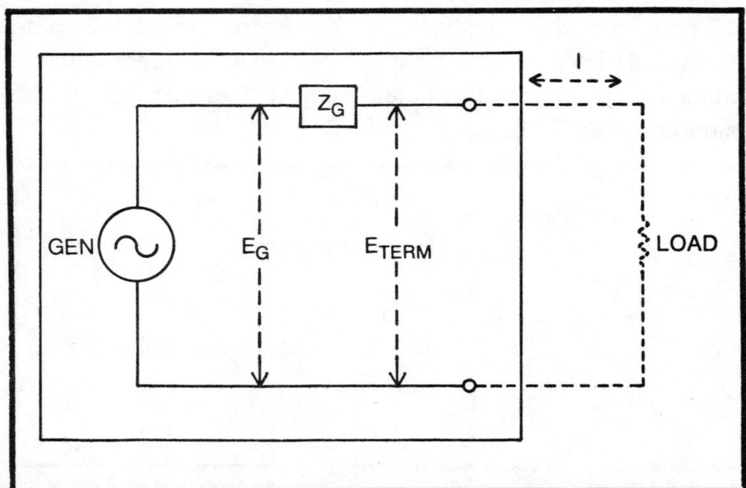

Fig. 2-9. Circuit illustrating impedance (Z_G) of an AC generator.

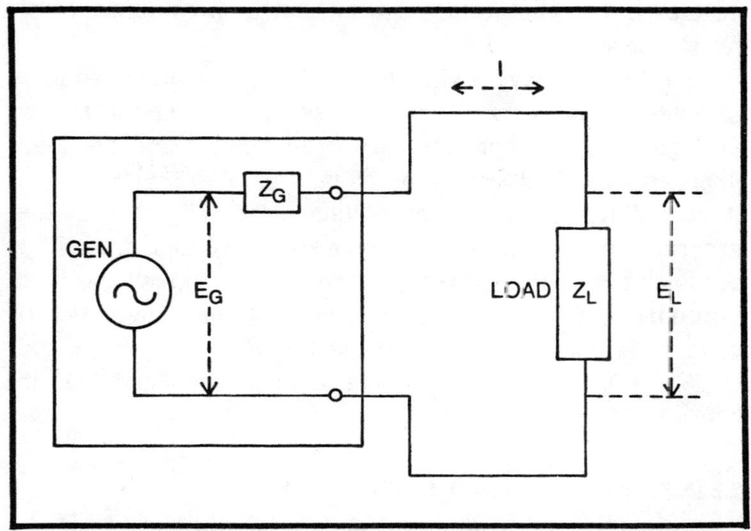

Fig. 2-10. Circuit illustrating an AC generator feeding a load impedance (Z_L). Current must flow through both the generator impedance (Z_G) and the load impedance.

generator voltage owing to the voltage drop across this impedance—$E_{TERM} = E_G - IZ_G$.

Like a mechanical generator, an electronic oscillator exhibits an internal impedance due to the output impedance of the tubes, transistors, or attenuators in the oscillator circuit. This internal impedance is generator impedance in the same sense as in the AC machine—since the oscillator is a nonmechanical producer of AC—but it is often called *oscillator output impedance*.

In practice, one may consider a generator to be any device or circuit that delivers a signal or power. This would include not only oscillators, multivibrators, machines, and other devices that form a signal, but also tubes and transistors and even any branch of a circuit that delivers a signal to another branch.

2.7 LOAD IMPEDANCE

Every AC load device has impedance (Z_L). Examples are loudspeakers, motors, lamps, heaters, transmitting antennas, etc. Sometimes, this impedance is resistive only; in other instances, it is classic impedance—that is, a combination of resistance and reactance.

Figure 2-10 shows a simple circuit in which an impedance Z_L loads an AC generator. In this setup, current must flow through the generator internal impedance Z_G and the load impedance Z_L. This current therefore is equal to $E_G/(Z_G + Z_L)$. It accordingly produces one voltage drop (IZ_G) across the generator internal impedance and a second voltage drop (IZ_L) across the load impedance. The voltage E_L across the load impedance (E_L) thus is somewhat less than the internal voltage of the generator, and (neglecting phase angle) is equal to $E_L = (E_G Z_L)/(Z_G + Z_L)$, where E is in volts and Z is in ohms.

2.8 INPUT AND OUTPUT IMPEDANCE

Every signal processing device or circuit, such as amplifiers, modulators, shapers, filters, etc., exhibits *input impedance* (Z_{IN}) seen by the applied signal and *output impedance* (Z_{OUT}) seen by the load device. These quantities must be dealt with in the design and application of the device, for the input driving-signal requirements, loading of the input-signal source, and load-device requirements depend upon Z_{IN} and Z_{OUT}.

Figure 2-11 illustrates the concept of a device having simple input and output impedances. In many instances, these

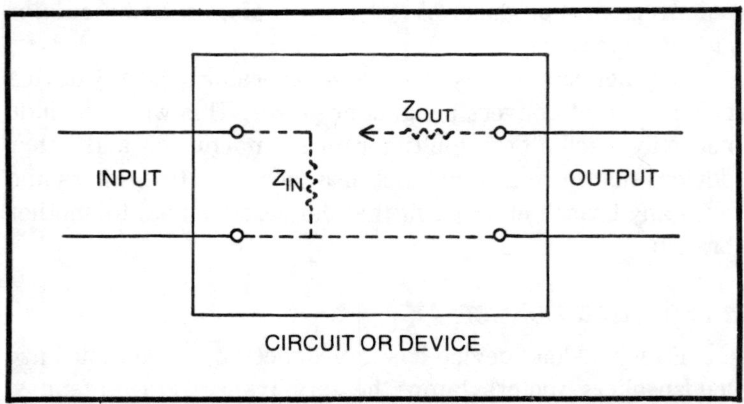

Fig. 2-11. Illustration of a device having both input and output impedance.

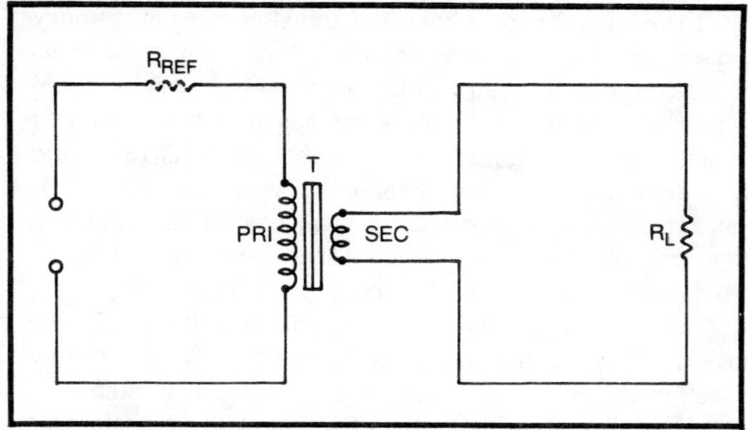

Fig. 2-12. Reflected impedance in the primary of a transformer with resistance R_L loading the secondary.

quantities are resistive. In most cases, the input impedance acts as a shunt component and the output impedance acts as a series component.

Some devices, such as amplifiers and filters, which receive and deliver signals, have both input and output impedance. Other devices, such as oscillators and transmitters, which are in effect generators, have only output impedance. Still other devices, such as meters and oscilloscopes, have only input impedance.

2.9 REFLECTED IMPEDANCE

When an impedance is connected across the secondary terminals of a transformer, a *reflection* of that impedance appears in the primary circuit of the transformer. This phenomenon is illustrated by Fig. 2-12; here, a resistance R_L loads the secondary. Because of R_L, an apparent resistance, called the *reflected resistance*, R_{REFL}, appears in the primary circuit.

The value of the reflected resistance depends upon R_L and the *turns ratio* of the transformer:

$$R_{REFL} = R_L(N_P/N_S)^2 \qquad (2\text{-}3)$$

where N_P is the number of primary turns, and N_S is the number of secondary turns.

It is not necessary to know the actual number of primary and secondary turns in order to use Eq. 2-13, but only the turns ratio as given by the transformer manufacturer or determined by user tests. Thus, a 3:1 transformer has three times as many secondary turns as primary turns; that is, $N_S/N_P = 3$, and $N_P/N_S = 0.3333$.* As an example, assuming an ideal transformer, if a 1000Ω resistor is connected to the secondary of a transformer having a 5:1 turns ratio, the reflected resistance at the primary terminals is: $R_{REFL} = 1000(1/5)^2 = 1000(0.2^2) = 1000(0.04) = 40\Omega$. If the turns ratio were 10:1, R_{REFL} would be 10Ω. With a stepup transformer, R_{REFL} is lower than R_L; with a stepdown transformer, R_{REFL} is higher than R_L; and, with a transformer having a 1:1 turns ratio, R_{REF} is is equal to R_L. These facts of performance lead to the general equation:

$$Z_{REFL} = Z_L(N_P/N_S)^2 \qquad (2\text{-}14)$$

Reflected impedance is of great importance in the technique of matching impedances by means of a transformer.

2.10 NEED TO MATCH IMPEDANCE

It is a fundamental axiom of electricity that maximum power is delivered by a generator to a load only when the load impedance equals the generator internal impedance. For this purpose, any device that delivers power can be considered a generator. The relationship is expressed:

$$Z_L = Z_G \text{ (for maximum power transfer)} \qquad (2\text{-}15)$$

Figure 2-13 illustrates this condition. In the circuit shown in Fig. 2-13(a), a variable load resistance (R_L) is connected to a 5V generator having an internal resistance (R_G) of 5Ω. Figure 2-13(b) shows the performance of the circuit as R_L is varied. From this table, note that as R_L is increased in 5Ω steps, from 5Ω to 50Ω, the total resistance ($R_G + R_L$) of the circuit increases from 30Ω to 75Ω; and the corresponding current, $I = E/(R_G + R_L)$, decreases from 0.167A when $R_L = 5\Omega$, to 0.067A when $R_L = 50\Omega$. Importantly, the power ($P = I^2R_L$) in the load increases from 0.139W when $R_L = 5\Omega$,

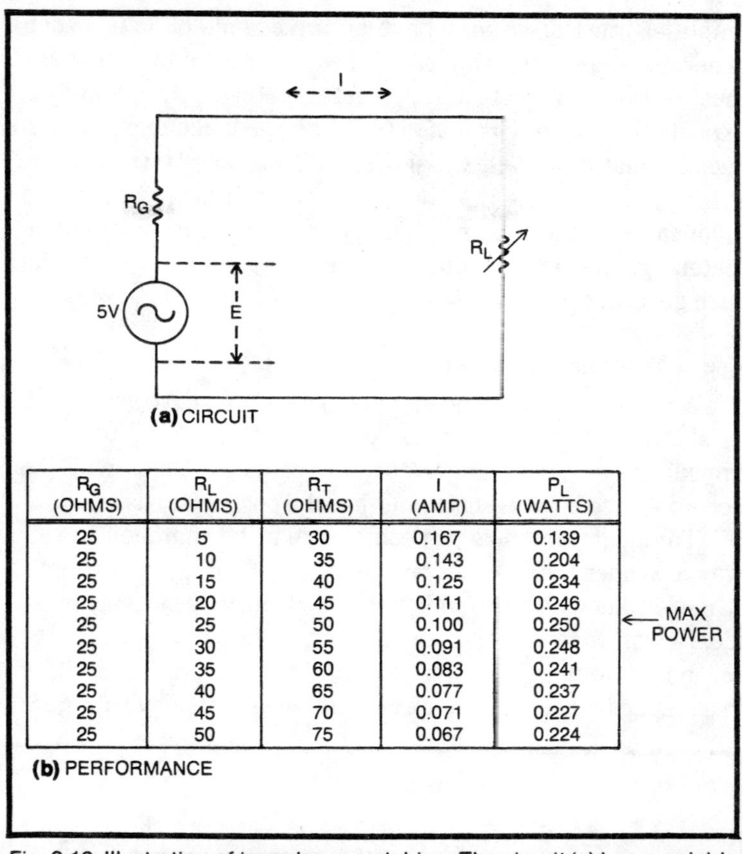

R_G (OHMS)	R_L (OHMS)	R_T (OHMS)	I (AMP)	P_L (WATTS)
25	5	30	0.167	0.139
25	10	35	0.143	0.204
25	15	40	0.125	0.234
25	20	45	0.111	0.246
25	25	50	0.100	0.250
25	30	55	0.091	0.248
25	35	60	0.083	0.241
25	40	65	0.077	0.237
25	45	70	0.071	0.227
25	50	75	0.067	0.224

Fig. 2-13. Illustration of impedance matching. The circuit (a) has a variable load resistance connected to a generator. The chart (b) shows the performance of the circuit as the resistance (R_L) is varied.

to 0.250W when $R_L = 25\Omega$; then, as R_L is further increased from 25Ω to 50Ω, the power decreases from 0.250W at 25Ω to 0.224W at 50Ω. The power peak thus is at 0.250W, the point at which $R_L = R_G = 25\Omega$.

2.11 METHODS OF MATCHING IMPEDANCE

For maximum power transfer, the impedance of a load device must equal that of the generator or other source. However, perfectly matched components are not always obtainable in practice. The output impedance of an amplifier, for example, may be 3500Ω, and the loudspeaker which the amplifier must drive may have an impedance of 3.2Ω. When generator impedance and load impedance do not match, steps must be taken to create a match between them. One technique exploits the phenomenon of reflected impedance explained in Sec. 2-9 and described under *Use of Matching Transformer*. Principal impedance-matching methods are described in the following subsections. In some areas, such as RF impedance matching, the representative method has been presented in each general category.

Use of Matching Transformer

A transformer may be inserted between a source and load, as shown in Fig. 2-14, for the purpose of matching the load impedance to the generator impedance. This will be accomplished if the transformer has the correct turns ratio.

To understand how impedances may be matched in this way, consider the instance in which R_L is a simple resistance. It is well known from fundamentals of electricity that, in an ideal transformer, the primary voltamperes equal the secondary voltamperes: $E_P I_P = E_S I_S$. This means simply that in a stepup transformer, for example, the secondary voltage is

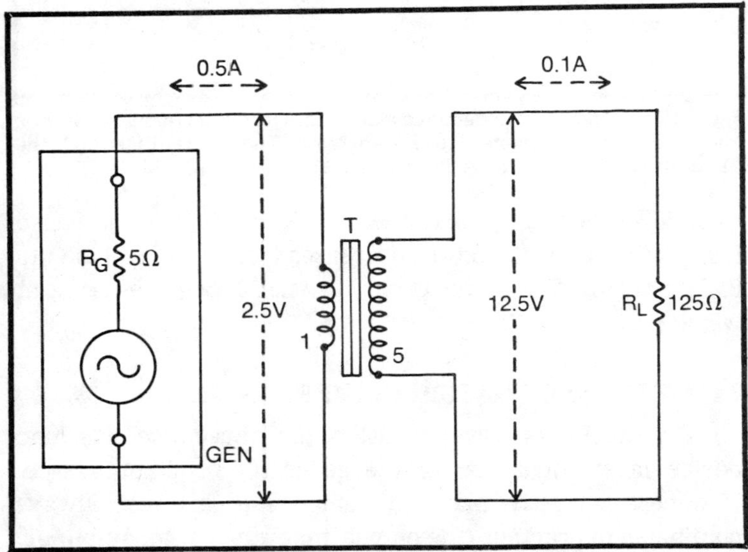

Fig. 2-14. Impedance-matching transformer.

higher than the primary voltage, but the secondary current is proportionately lower than the primary current, and that the opposite is true in a stepdown transformer. In Fig. 2-14, transformer T has a 5:1 stepup turns ratio. When the 5Ω generator (GEN) impresses 2.5V across the primary winding, 0.5A flows through the primary, and the primary voltamperes $= E_P I_P = 2.5(0.5) = 1.25$ VA. The 5:1 stepup gives a secondary voltage of 12.5V, and this forces a current of 0.1A through the 125Ω load resistor R_L. The secondary voltamperes is 12.5(0.1) = 1.25 VA, which is the same value as that of the primary voltamperes. Because the transformer has the correct turns ratio, it matches the 125Ω load to the 5Ω generator.

Observe that, although the turns ratio is 5:1, the impedance ratio is 25:1. Thus, the impedance ratio is the square of the turns ratio:

$$Z_S/Z_P = (N_S/N_P)^2 \tag{2-16}$$

And from this relationship, the necessary turns ratio for a required matching transformer is the square root of the impedance ratio:

$$N_S/N_P = \sqrt{Z_S/Z_P} \tag{2-17}$$

Example 2-10. A 2N3611 power transistor in the output stage of a 5W audio amplifier has a collector impedance of 20Ω. What turns ratio must an output transformer have to match this amplifier to a 3.2Ω loudspeaker?

Here, the impedance ratio $Z_S/Z_P = 3.2/20 = 0.16$. From Eq. 2-17, the turns ratio $N_S/N_P = \sqrt{0.16} = 0.4$, which indicates a stepdown transformer with a 0.4:1 turns ratio (the secondary has 0.4 of the turns in the primary). It should be noted, however, that impedance matching with a transformer involves working with the turns ratio and has nothing to do with the individual impedance of the primary and secondary windings.

The mention of a matching transformer usually brings to mind an iron-core device built for audio frequency use. It should be noted, however, that air-core transformers (tuned or untuned) are used in some instances for impedance matching at radio frequencies.

Use of Linear Devices

The impedance of linear devices, such as antennas and transmission lines, is described in Sec. 2.5, and equations for them are given there. The input, output, and characteristic impedances of some of these devices enable them to be employed for impedance matching at radio frequencies.

A common example is the matching of a transmission line to a transmitting antenna for the maximum transfer of energy from transmitter to antenna. In this application, the transmission line is termed a *feeder*. Figure 2-15(a) shows the connection of a coaxial feeder ($Z_0 = 72\Omega$) to the center of a half-wave antenna, where the antenna impedance approximates that of the feeder. At the transmitter end, the low-impedance feeder is matched to the impedance of the final amplifier by means of a small pickup coil (usually 1 to 3 turns) coupled to the amplifier tank coil, with the turns ratio providing the required impedance transfer. A twisted-pair transmission line sometimes is used in place of a coaxial feeder, but with greater losses.

Figure 2-15(b) shows how resonant open-wire feeders (the 600Ω type) are used to current-feed the center of the antenna. The center of the half-wave antenna is a high current point, and properly tuned quarter-wave feeders will have a high current at their antenna end. The length of feeders longer than a quarter-wavelength must so be chosen that a similar current loop occurs at the end; this requires that the feeder length be an even or odd multiple of a quarter-wavelength. Figure 2-15(c) shows how resonant open-wire feeders are used to voltage-feed a half-wave antenna by connecting them to one end of the antenna. Either end of the antenna is a high voltage point, and properly tuned quarter-wave feeders will have a high voltage point at their antenna end. As in the preceding case, the length of feeders longer than a quarter-wavelength must so be chosen that a similar voltage loop occurs at the end; this requires that the feeder length be an even or odd multiple of a quarter-wavelength.

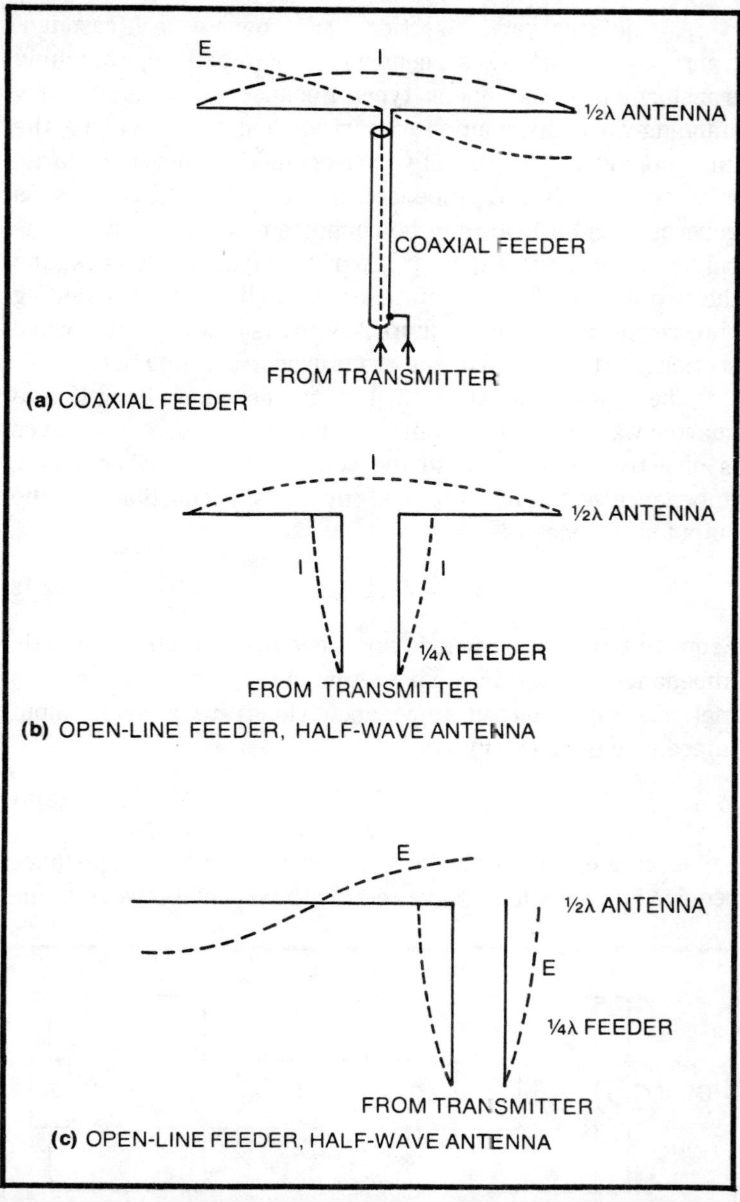

Fig. 2-15. Transmission line feeders used for the maximum transfer of energy from the transmitter to the antenna: (a) a coaxial feeder is connected to the center of a half-wave antenna; (b) a resonant open-line feeder is used to current-feed the center of a half-wave antenna; (c) a resonant open-line feeder is used to voltage-feed a half-wave antenna.

A quarter-wave section of open-wire resonant transmission line makes a convenient RF impedance-matching transformer of the linear type. Because of the stationary standing-wave distribution of current and voltage along the line, tapping into the line at various points can provide a large number of different impedances (see Fig. 2-7). Thus, a generator and a load may be connected, respectively, to the points corresponding to their separate impedance values, and the two devices become matches through the corresponding autotransformer action. Figure 2-16 shows how a quarter-wave section short-circuited at one end is used in this manner.

The input impedance (Z_{IN}) of a line whose length is a quarter-wave or an odd-numbered multiple of quarter-waves is directly proportional to the square of the characteristic impedance of the line (Z_0) and inversely proportional to the output impedance (Z_{OUT}):

$$Z_{IN} = Z_0^2/Z_{OUT} \qquad (2\text{-}18$$

From this relationship, it is apparent that the characteristic impedance a quarter-wave section must have, in order to match a given output impedance (load) to a given input impedance (generator), is:

$$Z_0 = \sqrt{Z_{IN}Z_{OUT}} \qquad (2\text{-}19)$$

Example 2-11. Calculate the characteristic impedance required for a quarter-wave section to be used between a line

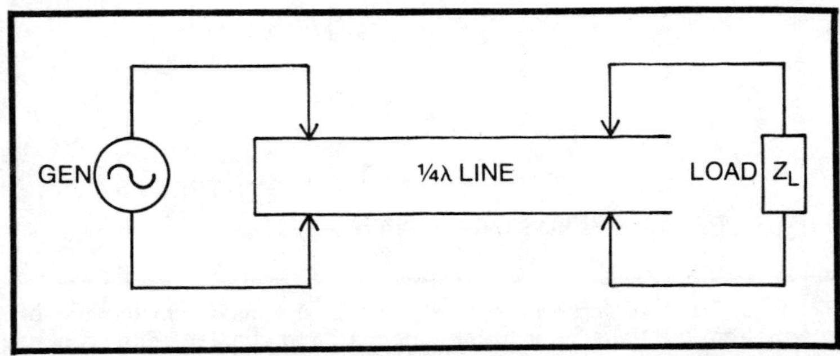

Fig. 2-16. A quarter-wave section of open-line resonant transmission line is short-circuited at one end and used as an RF impedance-matching transformer.

impedance (input) of 600Ω and an antenna impedance (output) of 72Ω.

From Eq. 2-19:

$$\begin{aligned} Z_0 &= \sqrt{600 \times 72} \\ &= \sqrt{43{,}200} \\ &= 207.85\Omega \end{aligned}$$

After finding Z_0 with Eq. 2-19, the required spacing of conductors in the section can be found with a rewritten form of Eq. 2-11:

$$S = \frac{d(\text{antilog } Z_0/276)}{2} \qquad (2\text{-}20)$$

Example 2-12. With No. 12 wires ($d = 0.081$ inch), the required spacing for the 207.8Ω is:

$$\begin{aligned} S &= \frac{0.081(\text{antilog } 207.8/276)}{2} \\ &= \frac{0.081 \text{ antilog } 0.752898}{2} \\ &= \frac{0.081(5.66106)}{2} \\ &= \frac{0.45854}{2} \\ &= 0.229 \text{ inch} \end{aligned}$$

Obviously, such close spacing of No. 12 wires (less than a quarter-inch) in a quarter-wave section would be impracticable in most instances. The remedy would be to increase the term *d* by moving to large-diameter conductors—such as metal rods or pipes. This results in the *Q-bar matching section* shown in Fig. 2-18(b) and described later.

A quarter-wave or half-wave section sometimes is used as an autotransformer to match a nonresonant feeder to an antenna as a load; in this application, the section is called a *matching stub*. Figure 2-17 illustrates this application, in (a) to centerfeed the antenna, and in (b) to endfeed it. In each instance, the stub is initially resonated by sliding the shorting bar to the proper point along the wires.

Other linear devices are similarly employed as RF transformers for matching nonresonant feeders to antennas. Two of these are shown in Fig. 2-18. In 2-18(a), the ends of the nonresonant feeder are flared out and attached to points equidistant from the center of the half-wave antenna. This matching section, called a *delta* (from its resemblance to the Greek letter Δ), provides a gradually increasing impedance. At a given operating frequency f, the delta dimensions are:

$$A = 118/f \qquad (2\text{-}21)$$

where A is in feet, and f in megahertz.

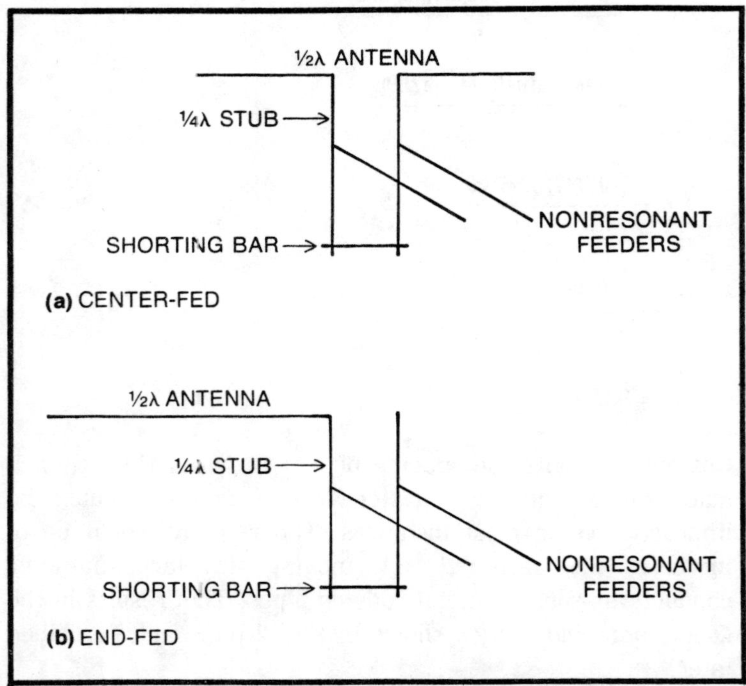

Fig. 2-17. Quarter-wave stubs are used to match nonresonant feeders to half-wave antennas: (a) center-fed, (b) end-fed.

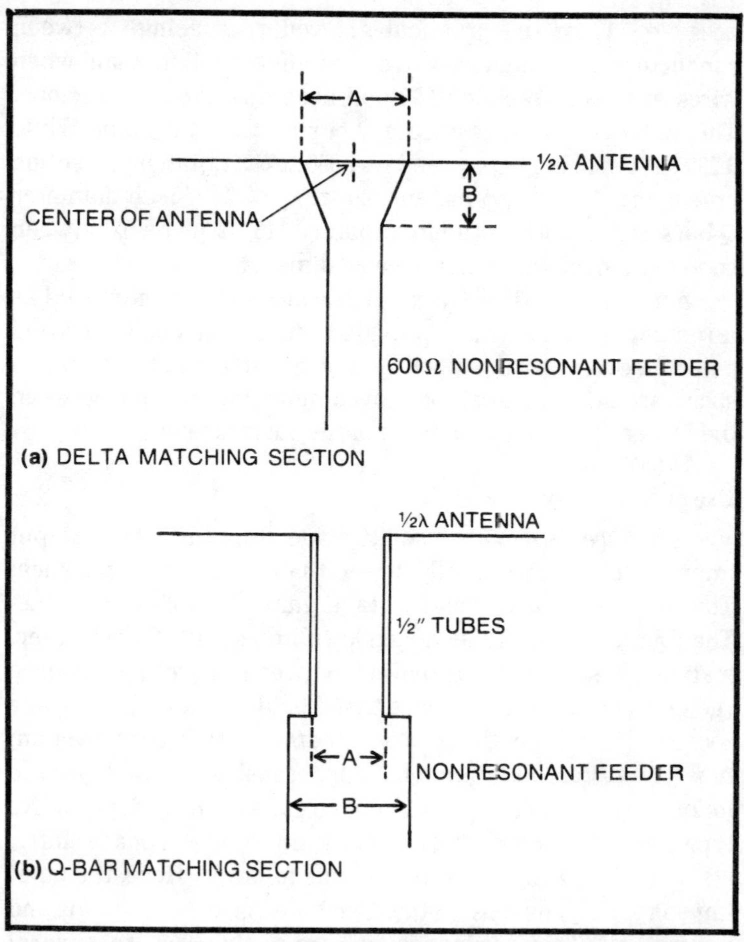

Fig. 2-18. RF impedance-matching transformers: (a) a delta matching section provides a gradually increasing impedance; (b) a Q-bar matching section provides more spacing between conductors.

And:

$$B = 148/f \qquad (2\text{-}22)$$

where B is in feet and f in megahertz.

In 2-18(b), a linear transformer consisting of two parallel lengths of half-inch-diameter aluminum tubing is connected between the nonresonant feeder and the center of the half-wave antenna. The large diameter of these tubes makes possible a more practicable, wider spacing between conductors in a quarter-wave matching section than when wires are used (see Eq. 2-20 and accompanying discussion). This arrangement is termed a *Q-bar matching section.* While 0.229 inch spacing is required in a 600-to-72Ω matching section employing No. 12 wires, the spacing of half-inch-diameter *Q*-bars is 1.41 inches (approximately 113/32 inches) between centers, a much more manageable dimension.

A suitable section of coaxial line also may be employed as a matching transformer, provided the center conductor and outer sleeve can be tapped at the correct points or that a short-circuiting disc can be moved along the interior between the center conductor and inside of the outer sleeve.

Use of Active Followers

A *follower* is usually a single-stage amplifier whose output impedance is substantially lower than its input impedance. The maximum theoretical voltage gain of a follower is one. The follower is useful as a stepdown impedance transformer, and often serves as a buffer between a voltage source (generator) and a load device that would overload the voltage source. There are three types: *cathode follower* (vacuum tube), *emitter follower* (bipolar transistor), and *source follower* (FET.) Figure 2-19 shows circuits of these devices. No type of follower, operating correctly, introduces a phase shift.

Figure 2-19(a) shows the cathode follower. Here, the input impedance equals closely the resistance of the grid-to-ground resistor r_G. This resistance is commonly 500K to several megohms, and can be made as high as desired, consistent with noise pickup and instability. The output impedance is:

$$Z_{OUT} = \frac{r_P r_K}{r_P + r_K\ (\mu + 1)} \qquad (2\text{-}23)$$

where Z_{OUT} is the output impedance in ohms, r_P the tube plate resistance in ohms, r_K the cathode resistance in ohms, and μ the tube amplification factor.

Example 2-13. A cathode follower employs a 6C4 tube having an amplification factor of 17, plate resistance of 7700Ω, and using a cathode resistor of 560Ω. Calculate the output impedance.

From Eq. 2-23:

$$Z_{OUT} = \frac{7700(560)}{7700 + (560 \times 18)}$$

$$= \frac{4,312,000}{7700 + 10,080}$$

$$= \frac{4,312,000}{17,780}$$

$$= 242.5\Omega$$

Figure 2-19(b) shows the emitter follower. In this circuit, unlike that of the cathode follower, the output impedance depends upon the source impedance Z_G:

$$Z_{OUT} = \frac{Z_G + h_{IE}}{1 + h_{FE}} \qquad (2\text{-}24)$$

where h_{IE} is the input impedance of the transistor in ohms, and h_{FE} is the forward-current transfer ratio of the transistor. Both h_{FE} and h_{IE} may be measured or taken from the transistor manufacturer's specifications.

Example 2-14. A type 40400 bipolar transistor has the following ratings: $h_{IE} = 600\Omega$ and $h_{FE} = 200$. Calculate the output impedance of an emitter follower employing this transistor with a 100K generator.

From Eq. 2-24:

$$Z_{OUT} = \frac{100,000 + 600}{1 + 200}$$

$$= 100,600/201$$
$$= 500.5\Omega$$

The input impedance of the emitter follower itself is equal approximately to $h_{IE} + h_{IE}R_E$, where R_E is the external emitter-to-ground resistor. In Fig. 2-19(b) and the preceding illustrative example, if R_E is 390Ω, a common value, then the

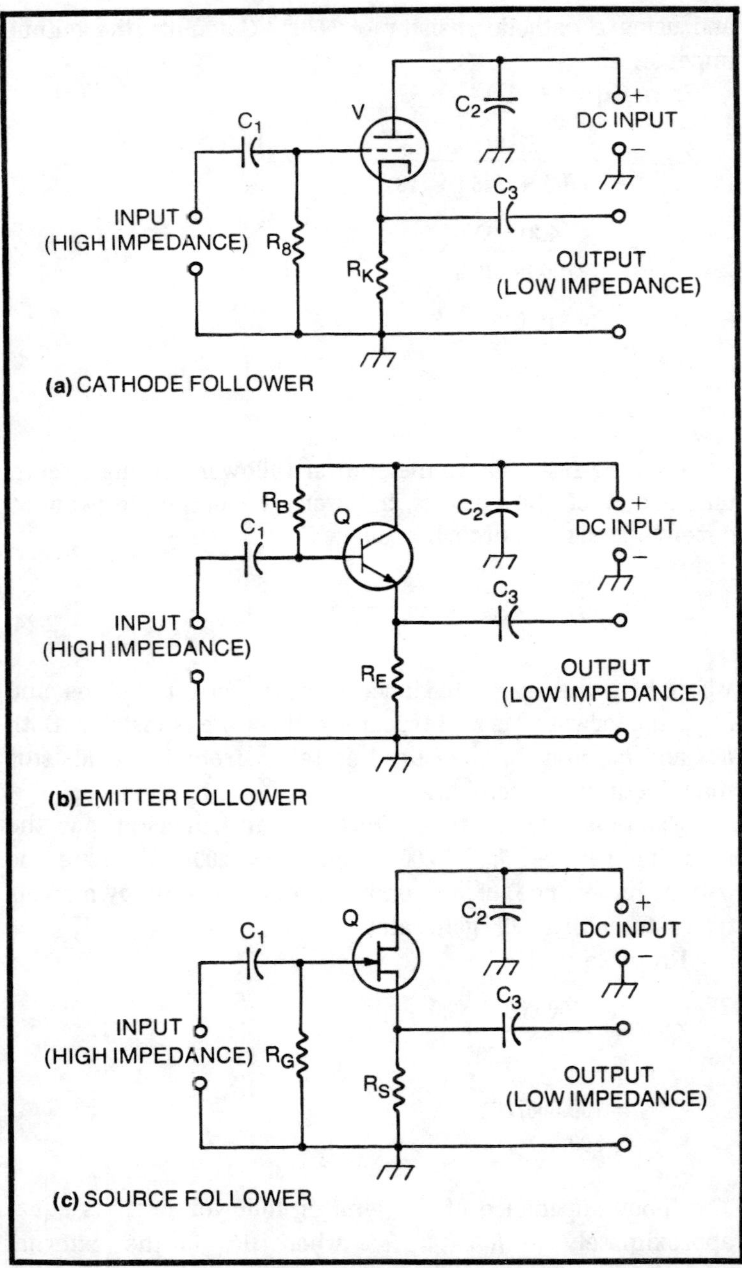

Fig. 2-19. Active follower circuits: (a) cathode follower; (b) emitter follower; (c) source follower.

input impedance is equal to: $600 + (200 \times 390) = 600 + 78{,}000 = 78{,}600\Omega = 78.6\text{K}$

The source follower in Fig. 2-19(c) behaves more nearly like the cathode follower. In the source follower circuit, the output impedance is:

$$Z_{OUT} = \frac{r_{OSS} r_S}{(g_{FS} r_{OSS} + 1) r_S + r_{OSS}} \tag{2-25}$$

where g_{FS} is the forward transconductance of the transistor in mhos, r_{OSS} the output resistance of the transistor in ohms, and r_S the resistance of the external source resistor in ohms. The input impedance is equal closely to the resistance of the gate-to-ground resistor, r_G.

Example 2-15. A type 40601 MOS field-effect transistor has a transconductance of 10,000 micromhos and an output resistance of 12K. Calculate the output impedance of a source follower employing this transistor with a 470Ω source resistor.

Here, $g_{FS} = 0.01$ mho, and $r_{OSS} = 12{,}000$ ohms. From Eq. 2-25,

$$Z_{OUT} = \frac{12{,}000 \times 470}{[(0.01 \times 12{,}000) + 1] \times 470 + 12{,}000}$$

$$= \frac{5{,}640{,}000}{[(120 + 1)470] + 12{,}000}$$

$$= \frac{5{,}640{,}000}{(121 \times 470) + 12{,}000}$$

$$= \frac{5{,}640{,}000}{56{,}870 + 12{,}000}$$

$$= \frac{5{,}640{,}000}{68{,}870}$$

$$= 81.9\Omega$$

Use of Pad-Type Attenuators

A *pad* consists of a combination of resistors so selected and arranged that the device, when inserted between a source and a load, presents a matching impedance to the source and load and introduces a desired amount of signal attenuation. The source and load impedances usually are resistive. Pads are named for the letters their arrangements resemble: L-*type*, T-*type*, and *pi-type*. Figures 2-20, 2-21, and 2-22 illustrate these three types and give the equations for determining the impedance values. From their shapes, the balanced-T is also called an H-*type*, and the balanced-pi an O-*type*.

In any proposed application of a pad, three factors are known preliminarily: the source (generator) impedance (Z_S), load impedance (Z_L), and desired attenuation (θ). The attenuation (loss) is expressed in nepers:

$$\theta = \text{dB}/8.686 \qquad (2\text{-}26)$$

where θ is the loss (attenuation) in nepers, and dB is the decibels ($= 10 \log_{10} P_1/P_2 = 20 \log_{10} E_1/E_2$, where P_1/P_2 output-to-input power ratio, and E_1/E_2 is the output-to-input voltage ratio).

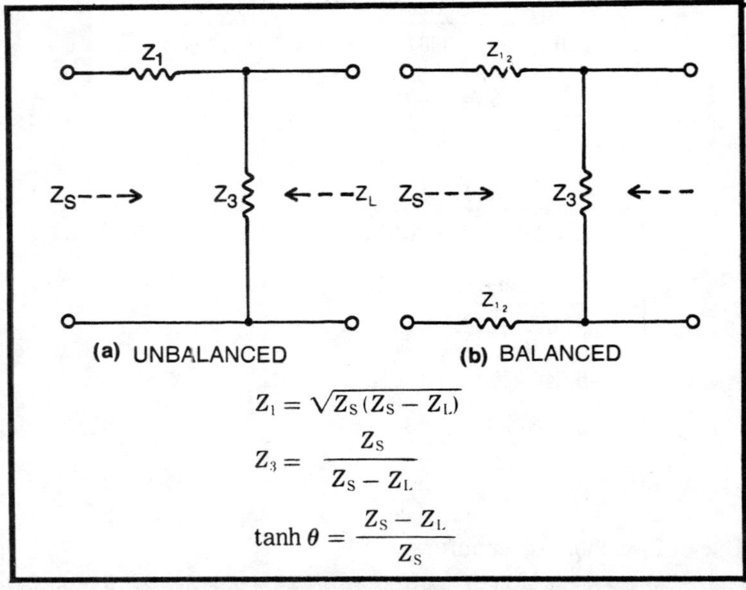

$$Z_1 = \sqrt{Z_S(Z_S - Z_L)}$$

$$Z_3 = \frac{Z_S}{Z_S - Z_L}$$

$$\tanh \theta = \frac{Z_S - Z_L}{Z_S}$$

Fig. 2-20. L-type pad attenuators: (a) unbalanced and (b) balanced. Equations given are for finding impedance values.

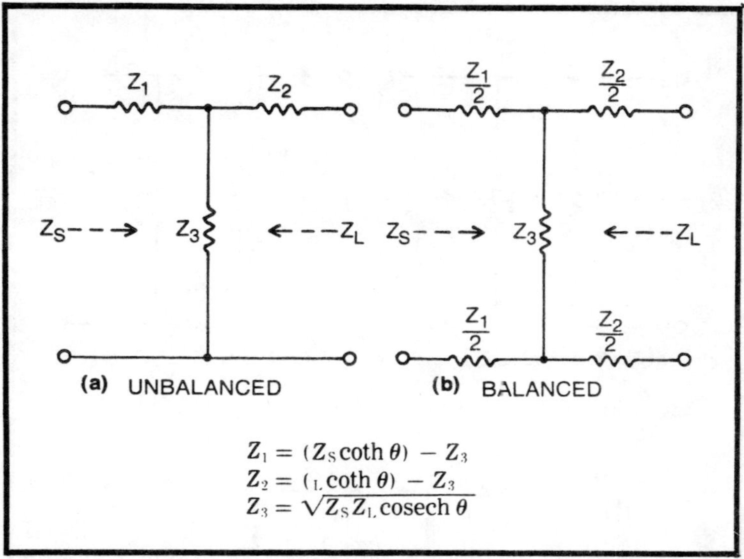

Fig. 2-21. T pad attenuators: (a) unbalanced and (b) balanced. Equations given are for finding impedance values.

From these factors, the resistances required in the pad are easily calculated. In each instance in Figs. 2-20 to 2-22, both balanced and unbalanced circuits are shown. The impedance values in each of the balanced circuits are determined from those calculated for the unbalanced circuit. The hyperbolic functions (sinh, tanh, coth, and cosech) may be obtained from a table of hyperbolic functions or by use of calculator which offers such functions on its keyboard.

In each case, the pad equations have been arranged in numerical order from Z_1 to Z_3; however, the user will find it advisable to solve for Z_3 first, since Z_1 and Z_2 both depend upon Z_3. In each circuit, Z_L, is connected to the output terminals. Likewise Z_L, the load impedance, is the impedance seen when looking into the output terminals of the pad when the source impedance Z_S is connected to the input terminals. For continuously variable attenuation, as in volume control, the resistances are ganged for simultaneous variation.

Example 2-16. A 10 dB unbalanced T pad is required to operate between a 500Ω source and 200Ω load. Calculate the required resistances. From Eq. 2-26, $\theta = 10/8.686 = 1.1513$.

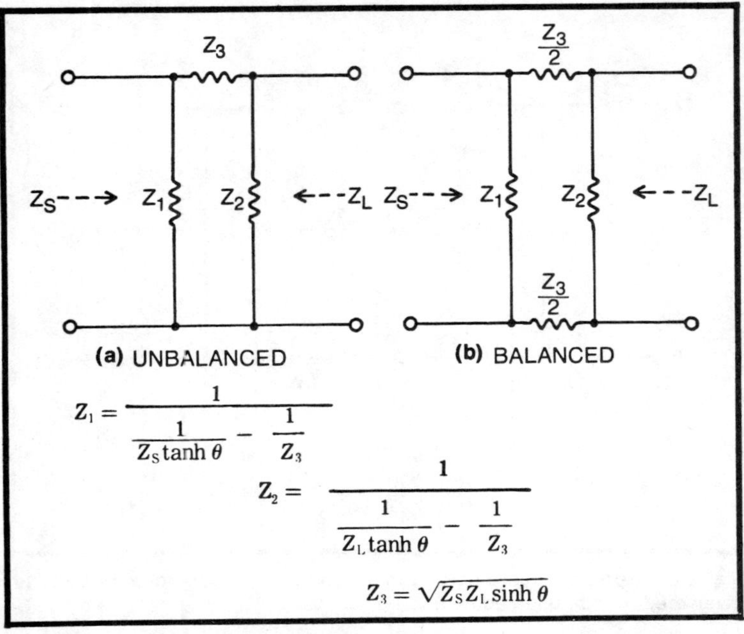

$$Z_1 = \frac{1}{\frac{1}{Z_S \tanh \theta} - \frac{1}{Z_3}}$$

$$Z_2 = \frac{1}{\frac{1}{Z_L \tanh \theta} - \frac{1}{Z_3}}$$

$$Z_3 = \sqrt{Z_S Z_L} \sinh \theta$$

Fig. 2-22. Pi-type pad attenuators: (a) unbalanced and (b) balanced. Equations given are for finding impedance values.

From function tables or calculator, coth $\theta = 1.2222$, and cosech $\theta = 0.70272$.

From Fig. 2-21:

$$\begin{aligned} Z_3 &= \sqrt{500(200)} \times 0.70272 \\ &= \sqrt{100{,}000} \times 0.70272 \\ &= 316.23 \times 0.70272 \\ &= 222.22\Omega \end{aligned}$$

And from Fig. 2-21:

$$\begin{aligned} Z_1 &= (500 \times 1.2222) - 222.22 \\ &= 611.1 - 222.22 \\ &= 388.88\Omega \end{aligned}$$

And from Fig. 2-21:

$$\begin{aligned} Z_2 &= (200 \times 1.2222) - 222.22 \\ &= 244.44 - 222.22 \\ &= 22.22\Omega \end{aligned}$$

2.12 ASPECTS OF IMPEDANCE

Certain evidences and concepts of impedance not touched upon in the preceding sections are described here. These topics are arranged alphabetically in the subsections below.

Common impedance. See mutual impedance—in a circuit.

Conjugate impedance is an impedance that has the same magnitude and the same resistance component as another impedance, but with a reactive component of opposite sign. Thus, if $X_L = X_C$ and $R_1 = R_2$, then impedances $Z_1 = \sqrt{R_1^2 + x_L^2}$ and $Z_2 = \sqrt{R2^2 + X_C}$ are equal. Impedance Z_2, having a negative capacitive reactance, is the conjugate or Z_1 whose inductive reactance is positive.

Driving-point impedance is the impedance that a network or device offers to the generator at the point at which the generator drives the network. Driving-point impedance is also termed *input impedance*.

Equivalent impedance is a term that has several meanings. In one sense, it denotes the equivalent series impedance of a parallel circuit—that is the impedance of a series circuit which, when connected to the same single-phase source, draws current of the same magnitude and phase angle as that drawn by the parallel circuit. In short, any impedance structure that can replace another structure without affecting current, voltage, and phase values is equivalent to the structure. For example, a certain T network may be equivalent to a certain pi network.

In another sense, an equivalent impedance (through simplification by means of Thevenin's or Norton's theorem) is the single impedance that corresponds to the combination of several others in a circuit. Impedances of the same kind combine in the same manner as resistors. Thus, for impedances in series, $Z_{EQ} = Z_1 + Z_2 + Z_3 + \ldots Z_N$. And for impedances in parallel, $Z_{EQ} = 1/(1/Z_1 + 1/Z_2 + 1/Z_3 + \ldots + 1/Z_N)$. This is also known as *total impedance* (Z_T).

Image impedance. For a network that connects a generator to a load, the image impedance (with respect to the load) is the total impedance of the generator and matching network, which is the same as the characteristic impedance of the generator. With respect to the generator, the image impedance is the total impedance of the matching network and load, which is the same as the characteristic impedance of the load. For example, with the proper load impedance attached to the network at one end, the generator sees an image impedance that is equal to the generator impedance; and, with the proper generator impedance attached, the load sees an image impedance that is equal to the load impedance.

Input impedance. See driving-point impedance.

Inverse impedance. See reciprocal impedance.

Mutual impedance. There are three common meanings of this term:

In a network. The mutual impedance is the apparent impedance ($Z = E/I$) between any two selected pairs of terminals of the network, with all other terminals open, where I is the current made to flow in through one pair, and E is the resulting open-circuit voltage across the other pair.

In a circuit. The mutual impedance is an impedance shared by two or more branches (sections or stages) of the circuit. Such an impedance—which may consist of a resistor, inductor, capacitor, or a combination of two or more of these—often causes signals to be miscoupled, either forward or backward, between sections or stages. An example of the trouble sometimes caused by such a mutual impedance is the motorboating in an amplifier, traceable to the common impedance of a power-supply output filter capacitor shared by several stages. This is also called *common impedance*.

Between neighboring antennas. A transmitting antenna (the "master") induces a voltage in any other nearby antenna (the "slave"). The mutual impedance between two such antennas is $Z_M = -E_2/I_1$, where I_1 is the current flowing at a selected point in the master antenna, and E_2 is the value of applied voltage that would be required at a selected point in the slave antenna (if the master antenna were not operating) to cause the flow of whatever current is observed in the slave as a result of excitation by the master.

Nonlinear impedance. Not every impedance obeys Ohm's law strictly; in some structures, current does not change linearly with a linear change in voltage. In some impedance devices, the entire response is nonlinear; in others, the nonlinearity occurs over only a part of the response curve. Nonlinear impedance is encountered in saturable reactors, tubes, transistors, semiconductor rectifiers, ceramic capacitors, voltage-dependent resistors, varactors, tungsten-filament lamps, and numerous other devices.

Poles of impedance are frequencies at which the driving-point impedance of a two-terminal reactive network is equal to infinity. (Compare *zeros of impedance*.)

Reciprocal impedances Two impedances (Z_1 and Z_2) are termed reciprocal to a third impedance (Z_3) when $Z_1Z_2 = Z_3^2$. Reciprocal impedances are also called *inverse impedances*.

Reciprocal of impedance The reciprocal of impedance is *admittance*, symbolized by Y and expressed in ohms: $Y = 1/Z$. The phase angle of an admittance vector is numerically equal to that of the impedance vector, but is of the opposite sign. Reciprocal impedance is also called *inverse impedance*, and has the same relationship to impedance that conductance has to resistance.

Total impedance in a two-mesh network is the quantity E_1/I_2, where E_1 is the voltage applied to the first mesh and I_2 is the resulting current in the second mesh.

Tube and transistor impedances are resistive internal impedances, and are governed by electrode direct currents and voltages. In tubes, examples are static plate impedance $z_P = E_P/i_P$, dynamic plate impedance ($z_P = de_P/di_P$, static screen impedance ($z_S = e_S/i_s$), dynamic screen impedance ($z_S = de_S/di_S$), static grid impedance ($z_G = e_G/i_G$), and dynamic grid impedance ($z_G = de_G/di_G$). A common example of working with these impedances is the matching of an amplitude modulator to an RF amplifier in a radio transmitter: a certain modulator tube draws a plate current of 500 mA at 500V ($z_P = e_P/i_P = 500/0.500 = 1000\Omega$), and the RF amplifier tube that it is to modulate draws a plate current of 250 mA at 1000V ($z_P = 10000.250 = 4000\Omega$). To couple the modulator to the amplifier for maximum power transfer, the required modulation transformer must have an impedance ratio of 1000Ω to 4000Ω, or 1:4 (see Sec. 2-11).

Except at high frequencies, the small internal capacitance and inductance of tubes does not significantly affect the impedance values. The input impedance, as determined by the internal capacitances of a vacuum tube whose grid is never driven positive, must sometimes be reckoned with in amplifiers operating at the upper end of the AF spectrum.

In the same way, impedances which are largely resistive are encountered in transistors, both bipolar and field-effect. Like corresponding tube impedances, these too are governed by electrode direct currents and voltages. Examples are static collector impedance ($z_C = v_C/i_C$), dynamic collector impedance ($z_C = dv_C/di_C$), static emitter impedance ($z_E = v_E/i_E$), dynamic emitter impedance ($z_E = dv_E/di_E$), static base impedance ($z_B = v_B/i_B$), dynamic base impedance ($z_B = dv_B/di_B$), and others.

Zeros of impedance are frequencies at which the driving-point impedance of a two-terminal reactive network is equal to zero (compare *poles of impedance*).

2.13 POWER FACTOR IN RELATION TO IMPEDANCE

For an AC circuit or device, the *power factor* is the ratio of power actually consumed (P) to the apparent power (VA = the simple product of volts and amperes): $pf = P/VA$. From this relationship, it can be seen that the maximum value which pF can have is one, and this would occur if, ideally, the power consumed equaled the simple calculated voltamperes. This condition can occur in a circuit or device containing pure resistance only ($P = VA$); but, in a practical AC circuit or device, resistance and reactance both are present, so pf has some value between one and zero. Thus, true power for the AC circuit or device is equal to $VA\,pf$ (or $EI\,pf$).

From basic electricity comes the simple formula for the power factor:

$$pf = \cos\theta \qquad (2\text{-}27)$$

where θ is the phase angle between current and voltage. The angle between current and voltage is the same angle between resistance and reactance (see Fig. 2-1). Therefore,

$$pf = R/Z = R/\sqrt{R^2 + X^2} \qquad (2\text{-}28)$$

which is identical to cos θ (see Fig. 2-1).

While the power factor is often expressed as a decimal in the manner just shown, it is sometimes expressed as a percent: *pf* 1 = 100%, *pf* 0.3 = 30%, etc.

Example 2-17. Calculate the power factor at 120 Hz of a filter choke having an inductance of 16H and a resistance of 580Ω.

Here, $X_L = 12{,}057.6\Omega$ (Eq. 1-10, Ch. 1), and Z (the impedance of the choke) = 12,071.5Ω (Eq. 2-2).

From Eq. 2-28:

$$\begin{aligned} pf &= 580/12{,}057.6 \\ &= 0.048 \end{aligned}$$

2.14 Q IN RELATION TO IMPEDANCE

Q is the *figure of merit* or *quality factor* of an AC device or circuit. It is the ratio of reactance to resistance:

$$Q = X/R \qquad (2\text{-}29)$$

where X and R are in ohms. From this relationship, $Q = \omega L/R$ and $Q = 1/(\omega CR)$. Q is also equal to tan θ (see Fig. 2-1) Q has no theoretical limit; if R were zero, Q would be infinite. In terms of impedance:

$$Q = \sqrt{Z^2 - R^2}/R \qquad (2\text{-}30)$$

Example 2-18. A certain 2.5 mH RF choke has a resistance of 125Ω and a 1 MHz impedance of 15.7K. Calculate the Q of this choke at 1 MHz.

From Eq. 2-3:

$$\begin{aligned} Q &= \sqrt{15{,}700^2 - 125^2}/125 \\ &= \sqrt{246{,}490{,}000 - 15{,}625}/125 \\ &= \sqrt{246{,}474{,}375}/125 \\ &= 15{,}699.5/125 \\ &= 125.6 \end{aligned}$$

2.14 PRACTICE EXERCISES

2.1. Calculate the impedance in ohms of a device that passes 60 mA for an applied voltage of 6.3V.

2.2. Calculate the impedance in ohms of a device that passes 30 μA for an applied voltage of 25 mV.
2.3. Calculate the voltage drop across a 50Ω impedance that carries 2 mA.
2.4. Calculate the voltage drop across a 3.2Ω impedance that carries 2A.
2.5. Calculate the current through a 2500Ω impedance for an applied test voltage of 1V.
2.6. Calculate the current through a 16Ω impedance for an applied voltage of 28.3V.
2.7. Convert 1380Ω to kilohms.
2.8. Convert 25,000Ω to gigohms.
2.9. Convert 580,000Ω to megohms.
2.10. Convert 0.5Ω to microhms.
2.11. Convert 0.1Ω to milliohms.
2.12. Convert 935,000Ω to teraohms.
2.13. Convert 1000K to gigohms.
2.14. Convert 500K to megohms.
2.16. Convert 0.05K to milliohms.
2.17. Convert 33K to ohms.
2.18. Convert 53,500K to teraohms.
2.19. Convert 5163 megohms to gigohms.
2.20. Convert 1000 megohms to kilohms.
2.21. Convert 0.01 megohm to microhms.
2.22. Convert 0.001 megohm to milliohms.
2.23. Convert 4.7 megohms to ohms.
2.24. Convert 50,000 megohms to teraohms.
2.25. Convert 1 gigohm to kilohms.
2.26. Convert 0.25 gigohm to megohms.
2.27. Convert 0.1 gigohm to microhms.
2.28. Convert 0.001 gigohm to milliohms.
2.29. Convert 2 gigohms to ohms.
2.30. Convert 0.3 gigohm to teraohms.
2.31. Convert 7 teraohms to gigohms.
2.32. Convert 15.2 teraohms to kilohms.
2.33. Convert 20 teraohms to megohms.
2.34. Convert 0.01 teraohm to microhms.
2.35. Convert 0.001 teraohm to milliohms.
2.36. Convert 0.8 teraohm to ohms.

2.37. Convert 1000 microhms to gigohms.

2.38. Convert 5520 microhms to kilohms.

2.39. Convert 10,000 microhms to megohms.

2.40. Convert 20 microhms to milliohms.

2.41. Convert 137 microhms to ohms.

2.42. Convert 15,500 microhms to teraohms.

2.43. Convert 35 milliohms to ohms.

2.44. Convert 1000 milliohms to kilohms.

2.45. Convert 150 milliohms to megohms.

2.46. Convert 10,000 milliohms to gigohms.

2.47. Convert 1,000,000 milliohms to teraohms.

2.48. Calculate the impedance offered by a 1000Ω resistance and a 2500Ω reactance in series.

2.49. Calculate the 400 Hz impedance offered by a 100Ω resistor and 100 mH inductor in series.

2.50. Calculate the 1000 Hz impedance offered by a 4700Ω resistor and 0.005 μF capacitor in series.

2.51. Calculate the impedance offered by 3900Ω resistance, 1000Ω inductive reactance, and 390Ω capacitive reactance in series.

2.52. Calculate the 500 Hz impedance offered by a 12H inductor, 0.01 μF capacitor, and 180Ω resistor in series.

2.53. Calculate the phase angle in degrees of a series circuit containing 1591Ω capacitive reactance and 1000Ω resistance.

2.54. Calculate the phase angle in degrees of a series circuit containing a 0.01 μF capacitor and 15K resistor and operated at 1000 Hz.

2.55. An accurate 0.005 μF capacitor is available. What value of resistance is required in series with this capacitance for a phase shift of 45 degrees at 2400 Hz?

2.56. A precision 1000Ω resistor is available. What value of capacitance in microfarads is required in series with this resistance for a phase shift of 60 degrees at 1000 Hz?

2.57. At what frequency in hertz will a series circuit of 0.5 μF and 9100Ω provide a phase shift of 75 degrees?

2.58. Calculate the phase angle in degrees of a series circuit containing 100Ω inductive reactance and 100Ω resistance.

2.59. Calculate the phase angle in degrees of a series circuit containing a 0.5H inductor and 1000Ω resistor and operated at 5 kHz.

2.60. Calculate the phase angle in radians of a series circuit containing a 10H inductor and 10K resistor and operated at 800 Hz.

2.61. What resistance is required in series with a 12H inductor to shift phase 45 degrees at 2 kHz?

2.62. At what frequency in hertz will a series circuit of 10 mH and 100Ω provide a phase shift of 30 degrees?

2.63. What inductance in millihenrys is required in series with 4700Ω to shift phase 40 degrees at 2400 Hz?

2.64. Calculate the impedance offered by 1000Ω resistance and 2500Ω reactance in parallel.

2.65. Calculate the 1000 Hz impedance offered by a 4700Ω resistor and 0.005 μF capacitor in parallel.

2.66. Calculate the phase angle in degrees of a parallel circuit containing 1591Ω capacitive reactance and 1000Ω resistance.

2.67. Calculate the phase angle in degrees of a parallel circuit containing a 5.5H inductor and a 2000Ω resistor and operated at 1000 Hz.

2.68. At what frequency in kilohertz will a parallel circuit of 0.002 μF and 1000Ω resistance provide a 45 degree phase shift?

2.69. Calculate the total impedance of the following similar impedances connected in series: 1000Ω, 800Ω, 350Ω, and 50Ω.

2.70. Calculate the equivalent impedance of the following similar impedances connected in parallel: 10,000Ω, 2250Ω, 1000Ω, 1000Ω, and 32Ω.

2.71. Calculate the *R* and *X* components of an impedance 150/65 degrees 15 minutes.

2.72. Calculate the *R* and *X* components of an impedance 16/45 degrees.

2.73. Calculate the impedance of a series circuit containing 1600Ω inductive reactance and 540Ω capacitive reactance.

2.74. Calculate the 400 Hz impedance of a series circuit containing 4 μF and 2.5H.

2.75. Calculate the 1000 Hz impedance of a series circuit containing 314Ω inductive reactance, 159Ω capacitive reactance, and 13.3Ω resistance.

2.76. Calculate the phase angle in degrees of the circuit in exercise 2.75.

2.77. Calculate the 2500 Hz impedance of a series circuit containing 30H, 0.01 μF, and 2700Ω in series.

2.78. Calculate the 500 Hz impedance in milliohms of a parallel circuit consisting of 1592Ω inductive reactance, 1000Ω capacitive reactance, and 3900Ω resistance.

2.79. Calculate the 120 Hz impedance in milliohms of a parallel circuit containing 30H, 2 μF, and 1000Ω in parallel.

2.80. Calculate the phase angle in degrees of a parallel circuit containing 240Ω inductive reactance, 1020Ω capacitive reactance, and 1000Ω resistance.

2.81. Calculate the characteristic impedance of a two-wire transmission line having a distributed capacitance of 50 pF and a distributed inductance of 100 μH.

2.82. Calculate the characteristic impedance of a two-wire transmission line made with No. 12 wire (diameter = 0.081 inch) with two-inch spacing.

2.83. Calculate the characteristic impedance of an air-insulated coaxial line in which the inner conductor has an outside diameter of 0.081 inch and the outer conductor has an inside diameter of 0.5 inch.

2.84. The dielectric constant of polyethylene is 2.3. If the transmission line in exercise 2.83 is filled with polyethylene, what will be its characteristic impedance?

2.85. A certain transformer has a turns ratio N_P/N_S of 0.5. Calculate the reflected impedance seen at the primary terminals when a resistance of 500Ω is connected to the secondary terminals.

2.86. What turns ratio is required in a transformer to match a 32Ω load to a 2500Ω source?

2.87. What impedance ratio is provided by a transformer having a turns ratio N_S/N_P of 10:1?

2.88. A certain multipurpose matching transformer has taps that provide turns ratios N_S/N_P of 2, 5, 20, 30, and 50 to 1. What impedance ratios does this transformer provide?

2.89. What characteristic impedance must a quarter-wave line have in order to match an output impedance of 300Ω to an input impedance of 75Ω?

2.90. If a certain quarter-wave line has a characteristic impedance of 300Ω, what output impedance will it match to an input impedance of 50Ω?

2.91. If a certain quarter-wave line has a characteristic impedance of 600Ω, what input impedance will it match to an output impedance of 1000Ω?

2.92. In a two-wire, 300Ω transmission line, what spacing in inches is required between two No. 12 wires (diameter = 0.081 inch)?

2.93. Calculate the horizontal (**X**) and vertical (**Y**) dimensions for a delta impedance-matching section to be operated at 14 MHz.

2.94. A type 8628 triode is used as a cathode follower with a 3300Ω cathode resistor. The plate resistance of this tube is 41,000Ω and the amplification factor is 127. Calculate the output impedance of the follower.

2.95. A type 2N3241A silicon transistor is used as an emitter follower driven by a 1000Ω signal source. For this particular transistor, $h_{FE} = 500$ and $h_{IE} = 700$. Claculate the output impedance of the follower.

2.96. A type 40603 MOS field-effect transistor is used as a source follower with a 270Ω source resistor. The output resistance (r_{OSS}) for this transistor is 4000Ω and the forward transconductance (g_M of g_{FS}) is 10,000 μmhos. Calculate the output impedance of the follower.

2.97. A 15 dB attenuation is introduced by a certain pad. Convert this figure to attenuation in nepers.

2.98. A three-neper loss is introduced by a certain pad. Convert this figure to decibels.

2.99. A certain pad has an input voltage of 5V and an output voltage of 1V. Express this loss in (a) decibels and (b) nepers.

2.100. From Fig. 2-20, Ch. 2, calculate Z_1 and Z_3 for an unbalanced-L pad to work between a 500Ω source and 100Ω load.

2.101. From Fig. 2-20, Ch. 2, calculate the resistance values for a balanced-L pad to work between a 500Ω source and 100Ω load.

2.102. From Fig. 2-21, Ch. 2, calculate Z_1, Z_2, and Z_3 for an unbalanced-T pad to work between a 1000Ω source and 150Ω load and provide 20 dB attenuation.

2.103. From Fig. 2-21, Ch. 2, calculate the resistance values for a balanced-T pad to work between a 1000Ω source and 150Ω load and provide 20 dB attenuation.

2.104. From Fig. 2-22, Ch. 2, calculate Z_1, Z_2, and Z_3 for an unbalanced pi pad to work between a 500Ω source and 200Ω load and provide 12 dB attenuation.

2.105. From Fig. 2-22, Ch. 2, calculate the resistance values for a balanced pi pad to work between a 500Ω source and 200Ω load and provide 12 dB attenuation.

2.106. A certain impedance $Z_1 = 500\Omega$ and another certain impedance $Z_2 = 1000\Omega$. To what third impedance Z_3 are these two impedances reciprocal?

2.107. What admittance Y corresponds to an impedance of 68.4Ω?

2.108. Convert 30 mhos admittance to impedance in milliohms.

2.109. A certain 8 μF capacitor has a power factor of 6% at 120 Hz. Calculate its equivalent series resistance.

2.110. The phase angle between current and voltage of a certain impedance device is 5 degrees. Calculate the power factor in percent of this device.

2.111. A certain 100 pF capacitor has a 1 MHz Q of 3000. Calculate the equivalent series resistance of this capacitor.

2.112. A certain 2.5 mH inductor has a resistance of 25Ω. What is the Q of this inductor at 500 kHz?

2.113. What value of impedance will the inductor in exercise 2.112 present at 500 kHz?

2.114. Calculate the 1 MHz impedance of a 50 pF capacitor having a Q of 1000.

(Correct answers are to be found in Appendix D.)

3 Impedance Measurements

Numerous methods are available for the measurement of impedance; some of these are direct and some are indirect. This chapter explains the techniques, providing a reasonable assortment to suit different conditions of instrument availability, operator experience, required frequency and impedance ranges, desired accuracy, and operator preference. Step-by-step instructions are given in most instances.

3.1 HINTS AND PRECAUTIONS

The measurement of impedance, like that of other electrical properties, is enhanced by the avoidance of pitfalls that can degrade a test. Detailed here are several areas in which technicians very often run into trouble.

Test Frequency

It is important that an impedance measurement be made at the proper frequency, for the Z value is different for each frequency even when the reactive component is small. There is no problem if the recommended operating frequency of a device or circuit is specified beforehand. When no frequency is given, however, the test frequency must be chosen on some logical basis; often, this will be the frequency at which the device will most probably be operated. In some instances, it is desirable to test a unit at several frequencies within a given operating range.

Common AF tests for impedance are 400 Hz and 1000 Hz. For power-supply components, 50, 60, 120, and 400 Hz are customary. Common RF tests (not including microwaves) are 100 kHz, 1 MHz, and 10 MHz. It is a mistake, of course, to assume—as some beginners do—that a simple 60 Hz test is satisfactory in all cases.

Waveform

A sinusoidal test signal must be used and the harmonic content of this signal must be as low as practicable (see Sec. 1.7, Ch. 1). A good quality signal generator supplies such a signal; but, even when such an instrument is used, the waveform should be inspected with an oscilloscope or distortion meter to insure that the test setup itself does not distort the signal.

A high harmonic content in the signal can cause meters to give false readings with the error sometimes being as high as the harmonic percentage.

Generator Impedance

The internal impedance of the test-signal source must be known, since it becomes a part of the measurement circuit, and must be accounted for in the calculation of impedance from current and voltage. While it is true that the output resistance of a signal generator is usually very low with respect to the impedance of devices that the generator normally drives, such as amplifiers and other high-impedance input circuits, this is not true in all impedance measurements. A signal generator with a 500Ω output, for example, might be called upon for checking impedances of 50Ω.

The impedance of most signal generators is resistive and is considered constant at all frequencies in the range of the instrument.

Instrument Impedance

The internal impedance of voltmeters and ammeters becomes a part of the test circuit and, as explained in individual tests in this chapter, must be accounted for in the calculation of impedance from current and voltage. Ideally, the impedance of a voltmeter is high to minimize current drawn by this instrument; and the impedance of an ammeter is low to minimize voltage drop introduced by this instrument. In electronic AC voltmeters and millivoltmeters, the internal resistance is 1–10 megohms on all ranges, depending upon make and model: and this is shunted by a capacitance between 20 pF and 40 pF. The resistance of rectifier-type nonelectronic voltmeters varies from as low as 1000Ω/V to 50K Ω/V, depending on make and model. The internal resistance of nonelectronic ammeters varies from 1400Ω for a 0.5 mA instrument to 1 mΩ for a 50A instrument. In transistorized electronic ammeters the internal resistance varies typically from 10K for the 10 μA range to 10Ω for the 10 mA range. A digital VOM on its alternating current ranges may present resistance varying from 1000Ω on the 200 μA range to 1Ω on the 200 mA range. Iron-vane AC ammeters have very low resistance, typically 213 milliohms for the 1A range to one milliohm for the 50A range. Because of their relatively high operating current, iron-vane voltmeters are not generally useful in impedance measurements. Iron-vane instruments are usually limited to operation at the power-line frequency.

Frequency Response of Instruments

Instruments used for impedance measurements must be accurate at the test frequency. High-grade laboratory-type AC voltmeter/millivoltmeter instruments of the electronic type retain their specified accuracy from five or ten hertz to upper limits of 100 kHz, 1 MHz, or 10 MHz, depending upon make and model. Special models may be employed, with suitable probes, for up to several gigahertz with reduced accuracy specified by the manufacturer. Kit-type electronic AC voltmeter/millivoltmeters usually are guaranteed up to one megahertz. Service-type VTVMs typically are rated from 50 Hz to 1 MHz or higher for AC voltage, depending upon make and model, and most are usable to higher frequencies with an RF probe. Service-type TVMs typically are rated from 50 Hz to 50 kHz for alternating voltage and current, depending upon make and model.

The frequency response of rectifier-type voltmeters and ammeters is poor for instruments equipped with a copper-oxide rectifier; in this type, a negative error usually appears at some point between 1 and 5 kHz and increases with frequency. Instruments equipped with point-contact rectifiers give better performance, often being usable up to 1 MHz. Since conventional (nonelectronic) VOMs employ rectifier-type meters, the frequency response of such multipurpose instruments depends upon the type of rectifier used.

Iron-vane and dynamometer-type instruments have a limited frequency range. The former are usually specified for 60 Hz operation, and the latter usually for a narrow band such as 25–125 Hz, 380–440 Hz, etc.

When impedance is to be measured at only one frequency, it is sufficient to know the accuracy of the instruments at that frequency alone. But when the measurements must be made at several frequencies, it is wise to examine the response of the instruments throughout the entire band.

Accuracy of Instruments

The accuracy of meters, bridges, and other instruments used in impedance measurements must be determined, and all impedances found from tests made with these instruments must be corrected accordingly. Depending upon make and model, the accuracy of analog-type AC voltmeters and ammeters ranges from 0.1% to ±5% of full-scale deflection, and the accuracy of the digital type ranges from ±0.5% (plus one digit) to ±1% (plus one digit) for voltage, and ±0.7% (plus one digit) to ±1% (plus one digit) for current. For best results, readings should be made in the upper quarter of the scale of an analog-type meter whenever possible.

Depending upon make and model, impedance bridges that separately evaluate resistance, capacitance, and inductance, from which impedance may be calculated, provide accuracy between ±0.05% and ±5% of the indicated value.

Operating Limits of Test Component

The test-signal voltage and current must be kept within the ratings of the impedance device which is under test. Not only will excessive voltage and current damage some components, but the response of some of them—such as iron-core inductors—becomes nonlinear when the current is too high, and the impedance under such conditions is *atypical*.

A general rule is to employ the minimum current and voltage that will give reliable results unless otherwise directed by the manufacturer of the component or the designer of the test.

Overdriving

Excessive test-signal amplitude is to be avoided. Not only is an overly intense signal liable to damage the component under test, but the distortion it sometimes produces can cause erroneous response of the instruments. Overdriving of some devices, such as amplifiers, can result in a false indication of input or output impedance.

Overloading

Overloading is the condition in which excessive current is drawn at some point in the test setup. A signal generator that is overloaded will sometimes cause erratic behavior of an impedance measuring circuit. A very common case of overloading occurs when the input impedance of a voltmeter in the test setup is too low; the meter accordingly draws excessive current and a false indication of test impedance may result.

Lead Length and Dress

At audio and high RF ranges, the most direct and shortest practicable leads must be used throughout an impedance measuring setup. Moreover, to minimize undesired coupling and capacitance, all leads must be kept as far apart as practicable.

External Fields

An impedance measuring setup must be protected from any interfering electric or magnetic fields. Often, this can be accomplished simply by moving all field-producing items—such as motors, generators, relays, power cords, transformers, chokes, etc.—from the vicinity of the setup, or by moving the setup to a field-free environment. In other instances, as in the pickup of radio stations by an RF impedance measuring setup, the setup itself must be adequately shielded.

Internal Fields

Sometimes interfering fields are produced within an impedance measuring setup itself. For example, the magnetic field of a power transformer or filter choke in a poorly shielded signal generator or other test instrument may cause trouble in the test circuit or may upset the operation of an unshielded meter. Also, the magnetic field of an inductor under test may penetrate other items, such as meters or coupling transformers, in the setup. The remedy is to make a preliminary "cleanup" test before any impedance measurements are attempted and correct any discovered faults.

Body Capacitance

In the low RF and high AF ranges, body capacitance—especially that associated with the operator's hands—can cause erroneous meter readings and sometimes frequency shifts. Often, the rearranging or shielding of components in an impedance measuring setup or the grounding of appropriate points in the circuit will correct this nuisance. Sometimes, however, it will be necessary to employ tuning wands to achieve distance between the operator and equipment. Each case is an individual one and no universal remedy is available. In stubborn cases when standard remedies are of no avail, a fixed relationship must be maintained between the equipment and the operator; that is, all adjustments must be made with the operator at the same distance and in the same position.

A particular nuisance is *antenna effect*, where the operator's body picks up signals from a strong radio station and couples them into the test setup. This condition usually is corrected by: choosing another, station-free, test frequency, if permissible; efficiently shielding the test setup; or removing the setup to a shielded room.

Temperature Effects

Most tests are made at room temperature (in the vicinity of 70°F), and the impedance values obtained at that temperature are acceptable unless the impedance must specifically be determined at some other point. Many test components are not especially temperature sensitive and their measured impedance does not change markedly if the ambient temperature fluctuates ten or twenty degrees in either direction. Some components, however, are temperature sensitive; these include thermistors, some resistors, capacitors, and RF inductors. These components should either be enclosed in a constant-temperature chamber during a test of their impedance, or they must be protected from hot resistors, transformers, and tubes in the test setup.

Vibration

Mechanical vibration, from whatever cause, is to be avoided in electronic measurements, but it is especially error-producing at RF and very high AF ranges where small displacements between components, caused by the vibrations, can be upset electrical relationships within the impedance measuring circuit. Vibration can also cause some meters to malfunction.

Resonance Effect

Electrical resonance may show up unexpectedly in an impedance measuring setup. Thus, an inductor under test may resonate with a coupling capacitor at the test frequency. Sometimes this is of no concern; at other times resonance may cause puzzling test results. Each case is individual and the operator must determine whether resonance is harmful and should be eliminated in a particular test of impedance.

Phase Relations

The operator should be aware of the various phase relations in a particular impedance measuring setup, particularly if tests are made in different branches of the circuit; otherwise, some perplexing situations may arise. If, for example, the circuit contains a capacitive reactance and an equal amount of resistance in series, the voltage across the capacitance (E_C) equals the voltage across the resistance (E_R); however, the total voltage across the circuit does not equal $E_C + E_R$, but is 1.414 E_C or 1.414 E_R because the phase angle here is 45 degrees. Similarly, if a voltage is applied to this circuit, neither the capacitor voltage nor the resistor voltage will be equal to half this value, but to $0.707E$. If a series-resonant circuit results from the connection of a test inductor in series with a coupling capacitor in the impedance measuring setup, the capacitor voltage (E_C) equals the inductor voltage (E_L), but the voltage across the circuit (generator voltage) will be much lower than either E_C or E_L.

Use of Same Instruments

Throughout an extended test, such as impedance measurements of the same component at a number of frequencies or bias voltages, the same instruments should be used whenever possible. If the use of different instruments is unavoidable, their characteristics should be carefully noted and any required corrections made to reconcile the results obtained with those obtained with the first instruments.

Often, when signal generators must be changed in order to provide the full required frequency range, the different output impedances of these instruments will cause trouble. Also, the accuracy of one generator may not be identical with that of another at overlapping frequencies.

3.2 VOLTMETER/AMMETER METHOD

A convenient and uncomplicated method of determining the value of an unknown impedance consists of passing a measured current through the impedance, measuring the resulting voltage drop across the impedance, and substituting the E and I values in the equation:

$$Z_X = E/I \tag{3-1}$$

where Z_X is in ohms, I in amperes, and E in volts.

Figure 3-1(a) shows the preferred test setup. In this arrangement, the internal impedance of the voltmeter (Z_M) must be much higher than the unknown impedance Z_X; otherwise the deflection of the ammeter will include both the current flowing through Z_X and the current taken by the voltmeter. When the voltmeter is an electronic instrument (VTVM or TVM), Z_M is several megohms and the current it demands is, to all practical intents and purposes, infinitesimal.

Test Procedure for Figure 3-1(a)

- Set up test circuit as shown in Fig. 3-1(a).
- Set generator to desired test frequency.
- Adjust generator output for ammeter deflection in upper quarter of scale and record deflection as current I in amperes.
- Read resulting deflection of voltmeter and record as E in volts.
- Using Eq. 3-1, calculate unknown impedance.

Example 3-1. When a certain device is tested in the circuit in Fig. 3-1(a) and the current is adjusted to 5 mA, the voltmeter reading is 3.2V. Calculate the unknown impedance.

Here, 5 mA = 0.005A. From Eq. 3-1,

$$\begin{aligned} Z_X &= 3.2/0.005 \\ &= 640\Omega \end{aligned}$$

When the voltmeter impedance is equal to or is less than Z_X, this meter cannot be used successfully to measure the voltage drop across Z_X and must be connected directly to the input of the test circuit, as shown in Fig. 3-1(b), to measure input voltage to the circuit. In this latter arrangement, ammeter $M2$ introduces a voltage drop so that the actual voltage E_3 across the unknown impedance Z_X is not equal to the

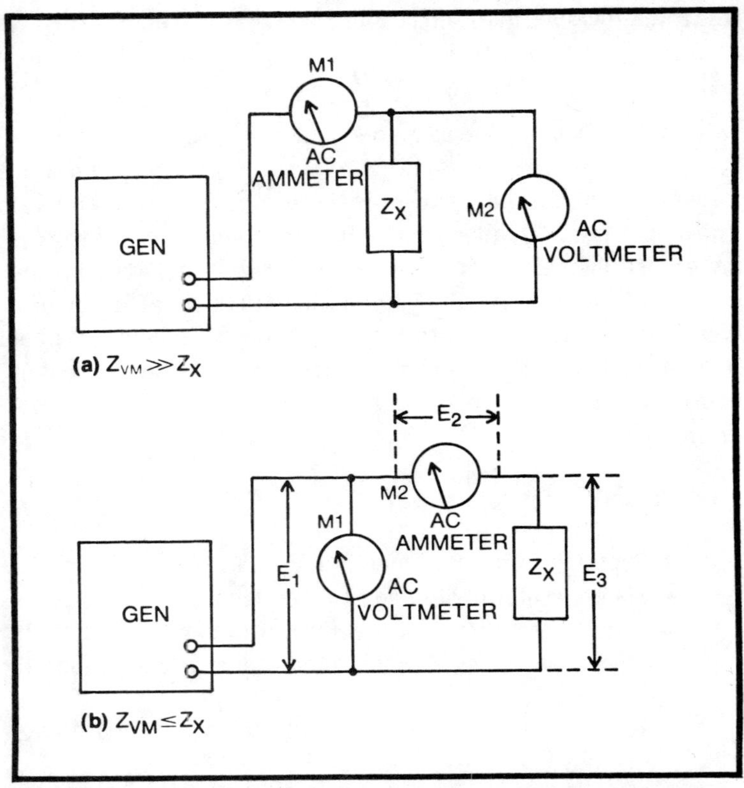

Fig. 3-1. Voltmeter/ammeter method for measuring an unknown impedance: (a) when the impedance of the meter (Z_{VM}) is higher than the unknown impedance (Z_X); (b) when the impedance of the meter is equal to or less than the unknown impedance.

voltmeter reading (E_1) but to E_1 minus the voltage drop E_2 across the ammeter. E_2 may be measured with a high-impedance voltmeter or it may be calculated:

$$E_2 = IR_M \tag{3-2}$$

where I is the indicated current in amperes, and R_M is the internal resistance of the meter (measured or taken from the manufacturer's literature).

For the test circuit in Fig. 3-1(b), the equation for unknown impedance becomes:

$$Z_X = (E_1 - E_2)/I \tag{3-3}$$

Test Procedure for Figure 3-1(b)

- Determine internal resistance of the ammeter and record as R_M in ohms.
- Set up test circuit as shown in Fig. 3-1(b).
- Set generator to desired test frequency.
- Adjust generator output for voltmeter deflection in upper quarter of scale and record as E_1 in volts.
- Read resulting deflection of ammeter and record as I in amperes.
- Using Eq. 3-2, calculate E_2.
- Using Eq. 3-3, calculate unknown impedance.

Example 3-2. A 0–5 mA AC meter in the test setup in Fig. 3-1(b) has an internal resistance R_M of 200Ω. When the generator output is adjusted for a voltmeter reading (E_1) of 10V, the indicated current is 4.5 mA. Calculate the unknown impedance.

Here, $I = 4.5\text{ mA} = 0.0045\text{A}$. From Eq. 3-2,

$E_2 = 0.0045(200) = 0.9\text{V}$. From Eq. 3-3:

$$Z_X = \frac{10 - 0.9}{0.0045}$$

$$= 9.1/0.0045$$

$$= 2022.2\Omega$$

If the voltage drop across $M2$ were ignored, and Eq. 3-1 used, the calculated impedance would be 2222.2Ω (a +9.9% error). When the unknown impedance is largely reactive, the Fig. 3-1(b) method becomes less reliable, since E_2 and E_3 are not then in phase with each other, and E_1 accordingly will not be equal to their sum (see item 17 in Sec. 3.1).

The voltmeter/ammeter method is widely used because most laboratories have the necessary meters, although they may not own other impedance measuring instruments. This method is usable equally well at the AF and RF ranges, provided the meters and generator have the required frequency capability and that care is taken in setting up and operating the test at high frequencies. (See items 5, 10, 13, 15, and 17 in Sec. 3.1.)

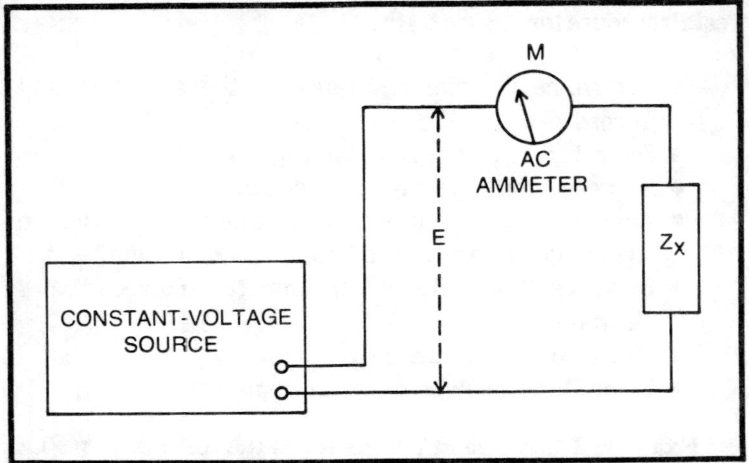

Fig. 3-2. Ammeter method for measuring an unknown impedance.

3.3 AMMETER METHOD

When a reliable source of constant-amplitude AC voltage is available to supply a single voltage (or several voltages in selectable steps), the voltmeter may be dispensed with in the impedance measuring setup described in the preceding section and only the current measured. This arrangement is shown in Fig. 3-2.

Here the known voltage is applied to the circuit, which consists of the current meter M (whose internal resistance R_M is known) and the unknown impedance Z_X in series, and the resulting current is read. From E and I, the unknown impedance then may be calculated:

$$Z_X = \frac{E - IR_M}{I} \qquad (3\text{-}4)$$

where Z_X is the unknown impedance in ohms, E the accurately known test voltage, I the indicated current in amperes, and R_M the internal resistance of the current meter in ohms.

Test Procedure

- Determine the internal resistance of ammeter and record as R_M in ohms.

- Set up test circuit as shown in Fig. 3-2.
- Apply test voltage E and record resulting deflection of current meter as I in amperes.
- Using Eq. 3-4, calculate unknown impedance.

Example 3-3. A regulated 10V AC source is used in the circuit shown in Fig. 3-2. The AC meter has a full-scale range of 1 mA and 0–1 AC an internal resistance of 600Ω. The total deflection is 0.75 mA. Calculate the unknown impedance in kilohms.

Here I = 0.75 mA = 0.00075A. From Eq. 3-4:

$$Z_X = \frac{10 \times (0.00075 \times 600)}{0.00075}$$

$$= \frac{10 - 0.45}{0.00075}$$

$$= 0.55/0.00075$$
$$= 12.733\Omega$$
$$= 12.733\text{K}$$

The ammeter method is simple, but its reliability depends upon the constancy of the voltage source. For continuous routine measurements of impedance, a direct-reading ohms scale may be drawn for the meter with its calibration being obtained from solutions of Eq. 3-4. The ammeter method may be used at the RF as well as the AF range, provided that the ammeter has the frequency capability and that necessary precautions are taken in wiring and operating the circuit (see items 5, 10, 13, 15, and 17 in Sec. 3-1). In fact, some operators employ the special constant 1V output of an RF signal generator for this test.

The ammeter method is susceptible to the effects of generator internal impedance Z_G. Since this impedance is in series with the unknown impedance and the ammeter, the current is proportional to the total impedance. Unless Z_G is very much smaller than Z_X (for example, $Z_X = 100\ Z_G$ or higher), Z_G must be subtracted from Eq. 3-4:

$$Z_X = \frac{E - IR_M}{I} \qquad (3\text{-}4)$$

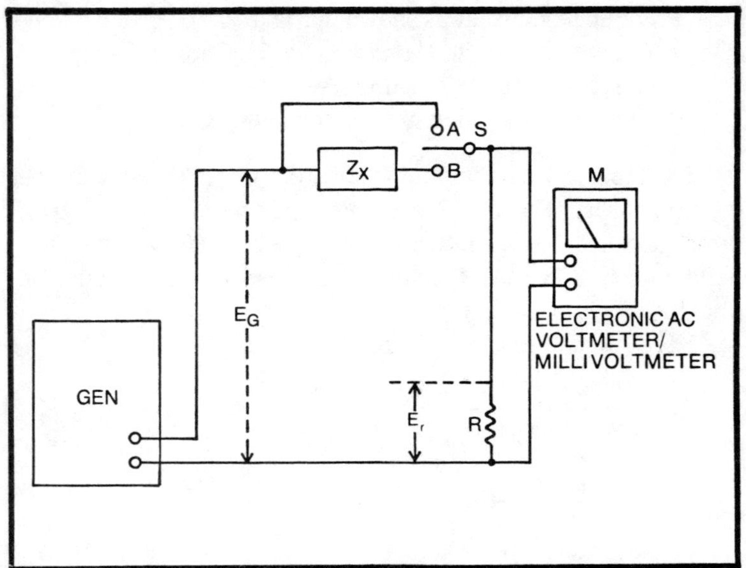

Fig. 3-3. Voltmeter method for measuring an unknown impedance.

3.4 VOLTMETER METHOD

An electronic AC voltmeter/millivoltmeter (VTVM or TVM) together with a standard resistor and changeover switch can be used to measure impedance over a wide range. Figure 3-3 shows the circuit.

In this arrangement, unknown impedance Z_X is connected in series with the generator and a low-resistance noninductive resistor R. (Common values used for the resistor are 1Ω and 10Ω.) The resistance is so low that current flowing through the resistor is determined by the impedance rather than by this resistance. Current flowing through the circuit sets up a voltage drop E_R across the resistor and this voltage is proportional to impedance Z_X. Switch S allows the meter to be switched to the input (position A) to read the applied test voltage E_G, or to the output (position B) to read the voltage drop E_r across the standard resistor. The unknown impedance is determined from these two voltages and the resistance:

$$Z_X = \frac{E_G R}{E_R} \tag{3-5}$$

where Z_X and R are in ohms, and E_G and E_R are in volts. When $R = 1\Omega$, Eq. 3-5 reduces to the simple ratio of the two voltages:

$$Z_X = E_G/E_R$$

where Z_X is in ohms and E_G and E_R are both in the same units (volts, millivolts, etc.).

Test Procedure

- Set up test circuit as shown in Fig. 3-3.
- Throw switch S to position B.
- Adjust generator output for an upper-scale deflection on selected scale of voltmeter M. Record reading as E_R.
- Without disturbing setting of generator, throw switch S to position A. Record new reading of voltmeter as E_G.
- Using Eq. 3-5, calculate unknown impedance from the two voltage readings and the resistance.

Example 3-4. A 10Ω resistor is used in the test setup in Fig. 3-3. The readings are $E_G = 1.5V$, and $E_R = 1$ mV. Calculate the unknown impedance in kilohms.

Here, $E_R = 1\text{ mV} = 0.001V$. From Eq. 3-5:

$$Z_X = \frac{10 \times 1.5}{0.001}$$

$$= 15/0.001$$
$$= 15{,}000\Omega$$
$$= 15K$$

This method is the most successful at the AF range. Because of feedthrough effects and stray reactances, it is difficult to use at frequencies beyond about 5 kHz.

3.5 SIMPLE, HOMEMADE, DIRECT-READING IMPEDANCE METERS

A self-contained impedance meter that reads directly in ohms may be made by calibrating any one of the circuits described in the preceding sections and providing a self-contained generator. Thus, in the circuit in Fig. 3-1(a), the scale of voltmeter M2 may be graduated in ohms based upon a selected value of current indicated by meter M1. Similarly, the scale of ammeter M2 in Fig. 3-1(b) may be graduated in ohms based upon a selected value of reference voltage indicated by meter M1.

In the circuit in Fig. 3-2, the scale of the ammeter may be graduated in ohms on the basis of the constant generator voltage E. The circuit in Fig. 3-3 would operate in a manner similar to that of a conventional ohmmeter. That is, with switch S set to position A, the generator output voltage would be initially adjusted for full-scale deflection of the meter (zero-impedance point); then, with s set to position A, a reading lower on the scale would indicate the impedance value. The graduations (obtained by calculation or by means of known impedances connected successively to the circuit) would extend from zero impedance at full-scale deflection to maximum impedance near zero deflection. A somewhat simpler method is to eliminate the switch and make the initial (zero) setting of the meter with the Z_X terminals temporarily short-circuited. This adaptation of the voltmeter circuit is subject to significant error, however, unless Z_X is largely resistive, since phase relationships between Z_X and R otherwise will cause a lower or higher reading than is anticipated (see item 17 in Sec. 3.1).

3.6 RESISTANCE/BALANCE METHOD

Figure 3-4 shows a circuit that can be used to measure impedances of all types. In this arrangement, the test signal is applied through transformer T to a variable resistor R and the unknown impedance Z_X in series. The same current flows through both R and Z_X; this current produces a voltage drop ($E_R = IR$) across the resistor and another voltage drop ($E_Z = IZ$) across the impedance. The electronic AC voltmeter M, either a VTVM or TVM, reads E_R when switch S is thrown to position A, and reads E_Z when S is thrown to position B. Transformer T serves only to isolate the generator from the circuit to prevent conflicting grounds between the generator and meter, so it need not be of any special type as long as it operates well at the test frequency. Performance of the circuit is based upon the fact that when the resistance is adjusted to the point that E_R equals E_Z, as noted by flipping switch *S* back and forth as *R* is adjusted, the resistance at that point equals the unknown impedance ($R = Z_X$). If the resistor is provided with a dial reading in ohms, the unknown impedance can be read directly from the dial; otherwise, the resistor may be disconnected from the circuit without disturbing its setting and checked with an ohmmeter or bridge. The maximum value of the variable resistor should equal the maximum impedance expected to be measured.

Test Procedure

- Set up test circuit as shown in Fig. 3-4.
- Throw switch *S* to position A.
- Set output of generator for suitable deflection of meter *M*. This delfection is voltage E_R. Do not subsequently disturb output of generator.
- Throw switch *S* to position B. Meter now reads E_Z. Observe difference between this voltage and value of E_R read in step 3.
- Throw switch *S* back and forth between positions A and B, while slowly adjusting variable resistor *R*, until

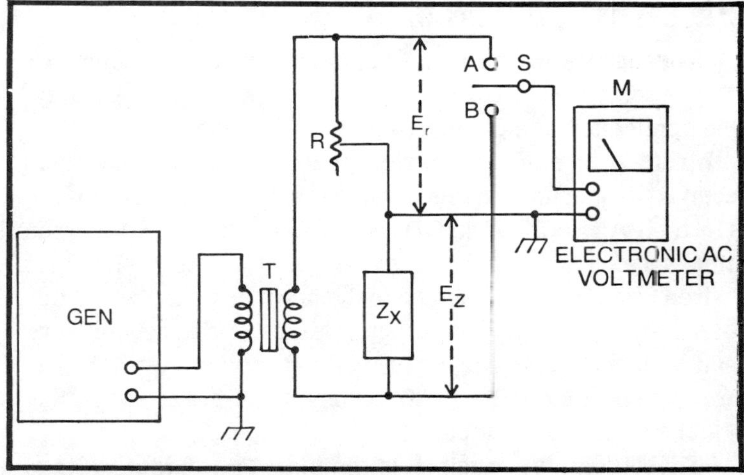

Fig. 3-4. Resistance/balance method of measuring an unknown impedance.

deflection of meter M is same at positions A and B. At this point $Z_X = R$ and can be read directly from resistor dial; or if the dial is uncalibrated, the resistor can be disconnected from the circuit and its setting checked with an ohmmeter or bridge.

The resistance/balance method is versatile. Its impedance accuracy corresponds to the accuracy with which the resistance is known. If a special dial is made for a one-turn potentiometer, the accuracy will coincide with that of the resistance-calibration source used; if a high-grade multiturn potentiometer is used with a turns-counting dial, without making an individual resistance calibration, an accuracy of 1% to 2% is possible. Resistance decade boxes need no calibration, since they automatically indicate their resistance settings. The accuracy of service-type decade boxes varies from ±2% to ±5%. Laboratory-type decade boxes afford accuracy as good as ±0.01%. Wirewound potentiometers, because of their inherent inductance, limit impedance measurements to the AF range; high-grade, frequency-compensated, laboratory-type decade boxes can be used up to 1 MHz or higher; however, in the RF range transformer T must be an air-core or ferrite-core unit.

3.7 SUBSTITUTION METHOD

Impedances of the same kind (capacitive or inductive) may be compared directly with the setup shown in Fig. 3-5(a). One application is a comparison of the impedance of devices with that of a standard device in line of production or in receiving inspection. In this arrangement, a simple T-network is formed by two 1K noninductive resistors (R_1 and R_2) and the impedance connected to terminals X–X. A test signal of desired frequency is presented to the network by the generator (GEN), and the input and output voltages of the network are read with the electronic AF/RF voltmeter. When switch S is thrown to position A the meter reads input voltage; when S is at B, the meter reads output voltage.

The operating principle is simple: The output voltage (E_{OUT}) is proportional to the impedance (Z_X) connected to terminals X–X and may be set to any selected value—the reference voltage—by adjusting the input voltage E_{IN} applied to the network with the impedance in place. The value of the reference voltage is not critical, so long as the value selected can be read accurately on the meter scale and can reset accurately by adjusting the input voltage. By varying the generator output, the output voltage is first adjusted to the chosen reference value E_0 with a known standard impedance (Z_S) in place. The corresponding input voltage is recorded as E_{I1}. Then, the unknown impedance is substituted for the standard impedance, and the input voltage is adjusted (by varying the generator output) to restore the output voltage to the original reference level. This new input voltage is recorded as E_{I2}. At this point, the unknown impedance may be calculated in terms of the known impedance:

$$Z_X = Z_S \frac{E_{I1} - E_{OUT}}{E_{I2} - E_{OUT}} \qquad (3\text{-}7)$$

where Z is in ohms, and all E's are in the same units (volts, millivolts, etc.). If the operator wants to know only how much larger or smaller Z_X is than Z_S—as in some forms of production testing—the desired multiplier M (whole number or decimal) may be calculated:

$$M = (E_{I1} - E_{OUT}) - (E_{I2} - E_{OUT}) \qquad (3\text{-}8)$$

Test Procedure

- Set up test circuit as shown in Fig. 3-5(a).
- Connect standard impedance Z_S to terminals X–X.
- Throw switch S to position B.
- Adjust generator output for a selected reference deflection of meter M (for example, 0.0V). Record this output reading as E_{OUT} and the corresponding input reading as E_{I1}.
- Remove Z_S and connect unknown impedance Z_X to terminals X–X.

- Readjust generator output voltage to restore meter reading to original E_{OUT} value.
- Throw switch S to position A. Meter now reads new input voltage E_{I2}.
- Using Eq. 3-7, calculate unknown impedance Z_X from E_{I1}, E_{I2}, and Z_S.
- If only the factor whereby Z_X differs from Z_S is required, use Eq. 3-8.

Example 3-5. In the test setup in Fig. 3-5(a), the circuit is initially adjusted with a 50Ω impedance (Z_S) in place and a reference voltage (E_0) of 0.1V. The input voltage (E_{I1}) at this point is 2.1V. When the unknown impedance (Z_X) is in place, and the test signal has been readjusted for the original E_0 of 0.1V, the new input voltage (E_{I2}) is found to be 1.77V. Calculate the unknown impedance.

Here, $E_0 = 0.1V$, $E_{I1} = 2.1V$, $E_{I2} = 1.77V$, and $Z_S = 50\Omega$. From Eq. 3-7,

$$\begin{aligned} Z_X &= 50[(2.1 - 0.1)/(1.77 - 0.1)] \\ &= 50(2/1.67) \\ &= 50(1.1976) \\ &= 59.9\Omega \end{aligned}$$

Example 3-6. The test setup in Fig. 3-5(a) is employed in an incoming inspection to check the deviation of the impedance of certain devices from the specified value of 135Ω. The reference voltage (E_0) is set to 0.5V with the standard impedance (Z_S) in place; the corresponding input voltage (E_{I1}) is 2.5V. Then, with one of the incoming devices (Z_X) in place, the input voltage must be set to 3V (E_{I2}). By what factor does Z_X differ from Z_S?

From Eq. 3-8, $M = (2.5 - 0.5)/(3 - 0.5) = 2/2.5 = 0.8$. Therefore:

$$Z_X = Z_S 0.8$$

Some AF and RF signal generators are equipped with output controls (or meters) which indicate directly the output voltage of the generator. When such a generator is available, the E_{I1} and E_{I2} values may be read directly from the generator and the switching arrangement shown in Fig. 3-5(a) can be omitted. This will result in the simplified circuit shown in Fig. 3-5(b). When this latter arrangement is used, all voltage readings—as before—must be in the same units: volts, millivolts, etc.

When an RF voltmeter is not available, a radio receiver having an internal S-meter may be used when checking RF impedance. A desired deflection of the S-meter (for example, center scale) may be selected as the E_0 reference point, but

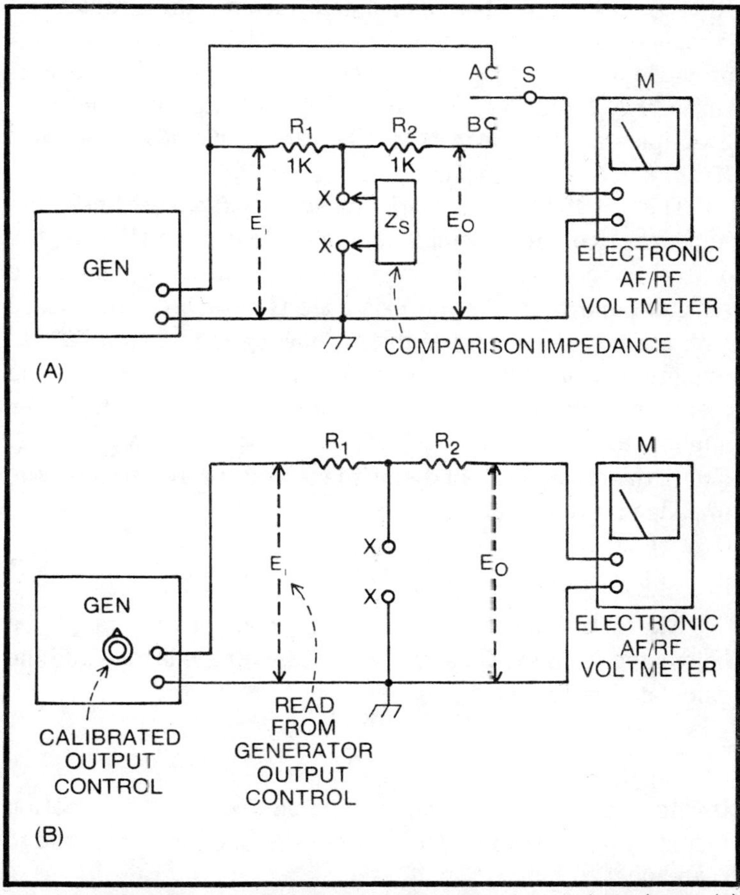

Fig. 3-5. Substitution method for measuring an unknown impedance: (a) to measure impedances of the same kind; (b) where voltage readings must be in the same units.

the actual corresponding RF voltage at the input of the receiver must be known and this voltage becomes the E_0 in Eq. 3-7 and 3-8.

3.8 IMPEDANCE BRIDGE

Most well-equipped laboratories have at least one impedance bridge (sometimes called a *universal bridge*). This is a multirange AC bridge for accurately measuring inductance, capacitance, AC and DC resistance, and loss factor (power factor, dissipation factor, Q, or all of these). Impedance may be calculated from the measured components. The impedance bridge usually operates at 1 kHz provided by a self-contained generator, or at other frequencies (usually 20 Hz to 20 kHz) provided by an external generator. Examples are the *Beco 307A; General Radio 1656; Heathkit IB-28;* and *Hewlett-Packard 4260A*.

This method is perhaps most satisfactory when the impedance contains a single dominant reactance (that is, the device is basically an inductor, capacitor, or inductive resistor) and when the AC resistance is separately measured with the bridge. When the test frequency is low—say 500 Hz maximum—the DC resistance may be used. From the measured value of capacitance C or inductance L, whichever applies, the reactance is calculated: $X_L = \omega L$, or $X_C = 1/\omega C$. From this reactance and the measured AC resistance, the impedance is calculated:

$$Z_X = \sqrt{R^2 + X^2} \qquad (3\text{-}9)$$

Alternatively, the resistance component may be calculated from the Q value measured with the bridge and the calculated reactance:

$$R_{AC} = X/Q \qquad (3\text{-}10)$$

Or the resistance may be calculated from the dissipation factor (D) measured with the bridge and the calculated reactance:

$$R_{AC} = DX \qquad (3\text{-}11)$$

Example 3-7. A certain choke coil which is to be used on AC only is checked with an impedance bridge at 1000 Hz and found to have an inductance of 1.6H and an AC resistance of 87Ω. Calculate the 1 kHz impedance of this choke.

Here, $X_L = 10{,}053\Omega$ (Eq. 1-10, Ch. 1), and $R = 87\Omega$. From Eq. 3-9:

$$\begin{aligned} Z_X &= \sqrt{87^2 + 10{,}053^2} \\ &= \sqrt{7569 + 101{,}062{,}809} \\ &= \sqrt{10{,}070{,}378} \\ &= 10{,}053\Omega \end{aligned}$$

Here, as might have been surmised, the resistive component is negligible compared with the reactive component. The 1000 Hz Q of this choke therefore is $Z/R = 10{,}053/87 = 115.5$.

Example 3-8. An electrolytic filter capacitor is checked at 120 Hz with an impedance bridge. The measured capacitance is 8.5 μF and the dissipation factor 0.057. Calculate the 120 Hz impedance of this capacitor.

Here, $X_C = 156\Omega$ (Eq. 1-12, Ch. 1). From Eq. 3-11, $R = 0.057(156) = 8.89\Omega$. And:

$$\begin{aligned} Z_X &= \sqrt{8.89^2 + 156^2} \\ &= \sqrt{79.03 + 24{,}336} \\ &= \sqrt{24{,}415} \\ &= 156.2\Omega \end{aligned}$$

It is interesting to note that the power factor of this capacitor (see Sec. 2.13, Ch. 2) is 5.69%, and that the impedance accordingly is only about 0.13% higher than the reactance of this capacitor at 120 Hz. At 1000 Hz the same capacitor might exhibit a power factor of 30% and an impedance of 19.1Ω, which is 11.6% higher than the reactance at 1000 Hz.

3.9 RADIO FREQUENCY BRIDGE

Whereas the impedance bridge is limited to audio frequency use, the *radio frequency bridge* is especially designed and constructed—with low-reactance resistors, adequate shielding, appropriate grounding, and other measures—for operation between 100 kHz and 250 MHz, depending upon make and model. Some RF bridges have a self-contained signal generator; some models require an external generator and null detector.

An example is the *General Radio 1606-B* RF bridge. This instrument has two calibrated balance controls, one for resistance and the other for reactance, and the dial of each reads directly in ohms. These balances are set separately for null in the same manner that the main balance control and power-factor control are set in the lower-frequency impedance bridge. From the measured R and X values, the unknown impedance may be calculated by means of Eq. 3-9. This instrument evaluates RF resistance between zero and 1000Ω, and reactance between −5000Ω and +5000Ω at 1 MHz. (At other frequencies, the dial reading at null is divided by the frequency in megahertz.)

It is convenient to be able to measure RF resistance directly, since this kind of resistance is quite complicated and may be significantly higher than either DC resistance or low-frequency AC resistance. Calculation of its value is

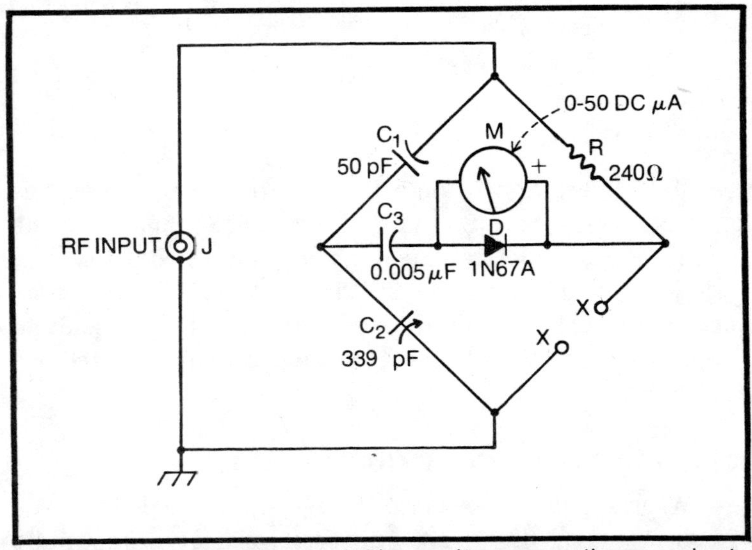

Fig. 3-6. A typical circuit for an RF bridge used to measure the approximate impedance of components used in transmitters, receivers, and antennas.

unreliable. Some of the factors that influence the value of RF resistance are skin effect, presence of dielectrics, presence of nearby conductors (such as metal shields), DC resistance, and stray reactance.

A special homemade version of the RF bridge is often used by radio amateurs for measuring the approximate impedance of components used in transmitters, receivers, and antennas. A typical circuit of this device is shown in Fig. 3-6. This arrangement is a four-arm bridge in which the arms are 50 pF fixed capacitor C_1, 339 pF variable capacitor C_2, 240Ω noninductive resistor R, and the unknown impedance Z_X connected to terminals X–X. The null detector is a simple RF voltmeter consisting of 1N67A germanium diode D, 0–50 DC microammeter M, and 0.005 μF coupling capacitor C_3. An RF voltage of approximately 1.75V_{RMS} is required for full-scale deflection of this meter. This bridge-driving signal is supplied by an RF oscillator or signal generator connected to coaxial input jack J. The unknown impedance is connected to terminals X–X by means of short, heavy, straight leads.

With the unknown impedance connected to terminals X–X, and an RF signal of desired frequency coupled into the bridge through input jack J, capacitor C_2 is tuned for null, as indicated by the lowest downward deflection of meter M. At that point, impedance Z_X is related to the 240Ω resistance of standard resistor R by the ratio C_1/C_2, where C_2 is the capacitance setting of variable capacitor C_2 at null. That is,

$$Z_X = R(C_1/C_2) \tag{3-12}$$

where Z_X and R are in ohms. and C_1 and C_2 are in picofarads.

If the dial of capacitor C_2 is a direct reading in ohms from a previous calibration of the bridge, the impedance may be read directly from the dial at null with no calculations being required. The simplest way to calibrate the dial is to connect a number of accurate noninductive resistors successively to terminals X–X, balance the bridge for each resistance, and inscribe that value on the dial. With the circuit constants shown in Fig. 3-6, variable capacitor C_2 will cover the impedance range of 35Ω to 600Ω. High impedances are at the low-capacitance end of the dial, and vice versa. Table 3-1 gives

Table 3-1. Sample Impedance/Capacitance Relationships.

C_2, including strays (picofarads)	Z_X (ohms)
344	35
240	50
120	100
80	150
60	200
40	300
24	500
20	600

a sample impedance-vs-capacitance relationship for the circuit. These values are based upon a basic tuning range of 14.7 pF to 339 pF (*Millen* 19335) and the knowledge that stray capacitance in the circuit adds at least 5 pF to the settings of the tuning capacitor. With solid construction and good shielding, the bridge is useful to 50 MHz.

3.10 Q-METER METHOD

RF impedance may be determined from *Q*-meter measurements. The procedure is to calculate reactance and resistance separately from the *Q*'s and tuning capacitances displayed by the *Q* meter, and then to calculate the impedance from *X* and *R*. When the usual parallel connection of the unknown impedance to the *Q*-measuring circuit is used, *X* and *R* are determined in the following manner:

$$X = \frac{1.1591 \times 10^8}{f \Delta C} \qquad (3\text{-}13)$$

where X is the reactance in ohms, f the test frequency in kilohertz, and ΔC the difference between C_1 and C_2. C_1 is the tuning capacitance in picofarads required to resonate the *Q*-measuring circuit *without* the unknown impedance connected, and C_2 is the tuning capacitance in picofarads required to reresonate the *Q*-measuring circuit *with* the unknown impedance connected. When $C_1 > C_2$, X is capacitive (−); when $C_1 < C_2$, X is inductive (+).

$$R = \frac{1.59 \times 10^8 C_1(Q_1 - Q_2)}{f(\Delta C)^2 Q_1 Q_2} \quad (3\text{-}14)$$

where ΔC and f are in the same units as in Eq. 3-13. R is the resistance in ohms. Q_1 the Q-meter reading when the Q-circuit is resonated *without* the unknown impedance connected. and Q_2 the Q-meter reading when the Q-circuit is reresonated *with* the unknown impedance connected. After X and R are determined as described above. the unknown RF impedance may be calculated with the aid of Eq. 3-9.

Example 3-9. A certain 100 pF capacitor is tested in a Q meter at 1 MHz. Without the test capacitor. the instrument is resonated with the tuning capacitor of the Q meter set to 400 pF (C_1). The Q reading (Q_1) at this point is 250. With the test capacitor connected. the instrument is reresonated with the tuning capacitor at 300 pF (C_2) and the corresponding Q reading (Q_2) is 75. From these C and Q readings the calculated value of Q_X is 26.78. Calculate the reactance X_C. resistance R. and 1 MHz impedance Z_X of the test capacitor.

Here. $f = 1$ MHz $= 1000$ kHz. $C_1 = 400$ pF. and $\Delta C = 400 - 300 = 100$ pF. From Eq. 3-13:

$$X_C = \frac{1.591 \times 10^8}{1000 \times 100}$$

$$= 1.591/100.000$$
$$= 1591\Omega$$

From Eq. 3-14:

$$R = \frac{1.591(10^8)400(250-75)}{1000(100^2)250(75)}$$

$$= \frac{1.591(10^8)400(175)}{1000(10.000)250(75)}$$

$$= \frac{1.1137 \times 10^{13}}{1.875 \times 10^{11}}$$

$$= 59.4\Omega$$

From Eq. 3-9:

$$
\begin{aligned}
Z &= \sqrt{59.4^2 + 1591^2} \\
&= \sqrt{3528.36 + 2531281} \\
&= \sqrt{2534809.36} \\
&= 1592.2\Omega
\end{aligned}
$$

The equivalent series resistance of 59.4Ω and the relatively low capacitor Q of 26.78 results in a 1 MHz impedance that is only about 0.07% higher than the reactance of this capacitor at that frequency.

3.11 USE OF TRANSMISSION LINE

RF impedance may be measured with a quarter-wave section of a two-wire transmission line, provided the measurement is made at the frequency at which the line is a quarter-wave long. (See *Transmission Lines* in Sec. 2.5, Ch. 2.) Figure 3-7 shows two practical ways of using the transmission-line method.

In Fig. 3-7(a) a noninductive resistor R is connected by the shortest practicable leads to one end of the line. Resistance R is equal to the characteristic impedance Z_0 of the line. The unknown impedance Z_X is connected by the shortest practicable leads to the other end of the line. An RF milliammeter (M1) is inserted in the line close to the transmitting end, and a second RF milliammeter (M2) is inserted in the line close to the receiving end. The internal resistance of the meters is very low. (R_M for a 0–115 RF milliammeter, for example, is approximately 5.5Ω, and for a 0–500 mA instrument is 0.63Ω.) The test signal from the generator (GEN) is loosely coupled into the line by means of a one-turn ring. The generator must be capable of supplying enough RF energy for an accurately readable deflection of the current meters. The unknown impedance is to the characteristic impedance of the line as the current at the sending end of the line (read by M1) is to the current I_Z at the receiving end of the line (read by M2): $Z_X/Z_0 = I_R/I_Z$. From this relationship, the unknown impedance can be calculated:

$$Z_X = Z_0(I_R/I_Z) \tag{3-15}$$

where Z_X is in ohms, and I_R and I_Z are in amperes.

Test Procedure

- Set up test circuit as shown in Fig. 3-7(a). Select resistance R equal to the characteristic impedance of the line.
- Set generator to a frequency corresponding to the wavelength at which the line is a quarter-wave.
- Adjust output of generator for accurately readable deflection of meters M1 and M2.
- Read current I_R from meter M1 and current I_Z from meter M2.
- Using Eq. 3-15, calculate the unknown impedance.

Example 3-10. An impedance device Z_X is connected to a quarter-wave 300Ω line in the setup shown in Fig. 3-7(a). Meter M1 reads 22.5 mA, and meter M2 5.3 mA. Calculate the unknown impedance.

Here, $Z_0 = 300\Omega$, $I_R = 22.5$ mA, and $I_Z = 5.3$ mA. From Eq. 3-15:

$$Z_X = 300 \frac{22.5}{5.3}$$

$$= 300 \times 4.24$$

$$= 1272\Omega$$

A test setup sometimes used by service technicians and radio amateurs to check the impedance of a quarter wavelength of transmission line is shown in Fig. 3-7(b). In this arrangement, a dip oscillator is inductively coupled loosely to the quarter-wave sample through a one-turn coil. The diameter of this coil should be about the same as that of the dip-oscillator coil.

With the receiving end of the line open, the oscillator is tuned downward throughout its frequency range. As this is done, several dip points (downward deflection of the meter) will be noticed. The lowest dip point occurs when the oscillator is tuned to the frequency at which the line is a quarter wavelength. At this point, a noninductive variable resistor R is connected to the line and adjusted to the point at which the dip disappears. The resistance at this setting equals the

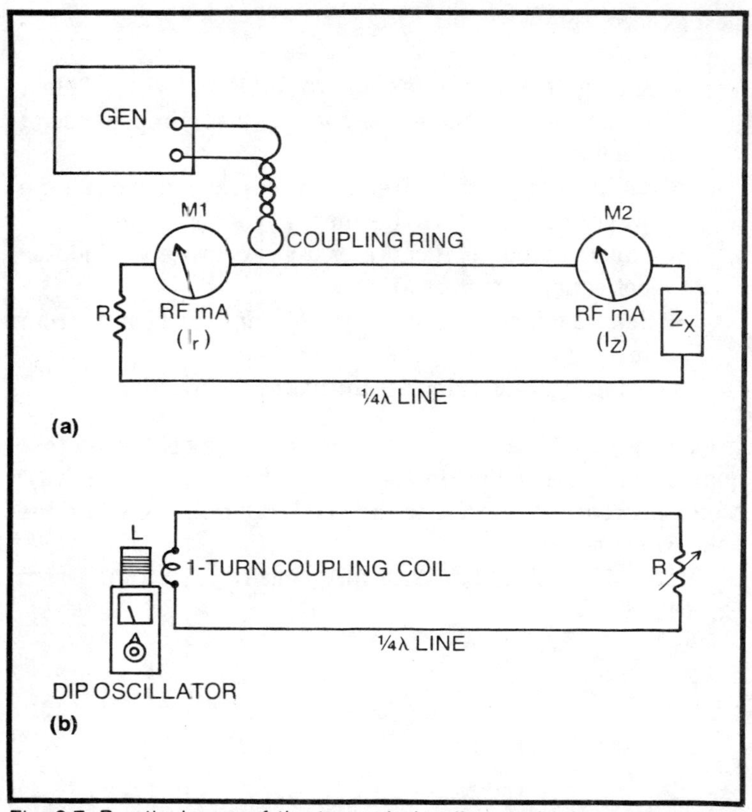

Fig. 3-7. Practical uses of the transmission line method to measure impedance: (a) using a noninductive resistor; (b) using a diposcillator.

characteristic impedance of the line ($Z_0 = R$) and can be read directly from the resistor dial if the latter has previously been calibrated in ohms. If the resistor dial is not calibrated, it may be disconnected and its resistance setting checked with an ohmmeter or bridge.

This is a convenient test, but it requires a rheostat or potentiometer that will operate efficiently at the radio frequency employed. Since this resistor must be noninductive, the only choice for many users will be a small composition potentiometer of the volume-control type which is not necessarily designed for RF use. Even when solid construction and careful operating techniques are employed, the accuracy of measurements may be expected to decrease as the frequency increases. (For general precautions, see items 5, 10, and 13 in Sec. 3.1).

The dip-oscillator method is often used to check the impedance of insulated transmission lines such as coaxial cable and TV ribbon. In this use, however, it must be remembered that a yardstick-measured length of such insulated line may not be an *electrical* quarter-wave long, its actual electrical length being longer and dependent upon the kind of dielectric used. For insulated line, a quarter-wave (in ft) = $246V/f$, where f is the test frequency in megahertz and V is the velocity factor for the particular kind of line (obtained from the line manufacturer's literature).

The utility of the transmission-line method, as illustrated by Fig. 3-7, is limited by the length of line that can be handled comfortably. Experience shows that the longest which can be worked with under ordinary conditions is about 12 ft, and this would correspond to a quarter wavelength at about 20 MHz. Also, the shortest length would not be much under a foot (a 9.5-inch line is a quarter-wave at 300 MHz). Therefore, the method appears to be limited in practice to the frequency range of 20 MHz to 300 MHz. Table 3-2 lists quarter wavelengths of line required at common frequencies between 20 and 300 MHz. In case a special line must be constructed for impedance measurements, Table 3-3 shows the spacing of two No. 12 wires (in inches) required for four common impedances.

Table 3-2. Quarterwavelengths of Open-Wire Transmission Line.

FREQUENCY (megahertz)	LENGTH OF ¼ WAVE
20	11′11″
30	7′11½″
40	5′11½″
50	4′9¼″
60	3′11-¾″
100	2′4-¾″
200	1′2¼″
300	0′9½″

Table 3-3. Spacing of No. 12 Wire.

IMPEDANCE Z_0 (ohms)	CENTER-TO-CENTER SPACING (inches)
200	0.215
300	0.495
500	2.62
600	6

3.12 USE OF SLOTTED LINE

At microwave frequencies, impedance may be measured indirectly by use of a slotted line. Figure 3-8 shows the test setup. This is a conventional arrangement: The microwave generator supplies RF energy to the slotted line at the desired test frequency, and the unknown impedance Z_X is connected to the opposite end of the line. The carriage is slid along to locate maximum-voltage points (maxima or *loops*) and minimum-voltage points (minima or *nodes*) and these points are indicated by maximum and minimum deflections of the meter in the detector. The slotted line has a characteristic impedance Z_0 (for example, 50Ω) specified by its manufacturer; and when the line is terminated in this impedance (that is $Z_X = Z_0$) there are no standing waves and the meter gives a steady deflection as the carriage is moved along. When Z_X is some value other than Z_0, loops and nodes are detected, and the unknown impedance is determined from the maximum and minimum values and the characteristic impedance:

$$Z_X = Z_0(E_{MAX}/E_{MIN}) \qquad (3\text{-}16)$$

where Z_X is the unknown impedance in ohms, Z_0 the characteristic impedance of the line in ohms, E_{MAX} the loop voltage, and E_{MIN} the node voltage. It is not mandatory that E_{MAX} and E_{MIN} be in volts, so long as they both are in the same units (volts, millivolts, microvolts). Some detectors read in current units and for them the multiplier in Eq. 3-16 would be I_{MAX}/I_{MIN}. Still other detectors read in arbitrary units, and the ratio would be a simple quotient of the two numerical readings.

Test Procedure

- Set up a test circuit as shown in Fig. 3-8 with an unknown impedance connected to the receiving end of the slotted line.
- With generator and detector operating, slide the carriage along to the point at which upward deflection of the meter indicates a loop (voltage or current maximum). Adjust output of generator to place this deflection at or near full scale. Record peak deflection as E_{MAX}.
- Slide carriage along to adjacent point at which a downward dip of the meter indicates a node (voltage or current minimum). Record bottom of deflection as E_{MIN}.
- Using Eq. 3-16, calculate unknown impedance.

Example 3-11. A 50Ω slotted line is used in the setup shown in Fig. 3-8. At a loop the voltage is set (by adjusting the generator output) to 10 mV. At the adjacent node the voltage is 3 mV. Calculate the unknown impedance.

From Eq. 3-16:

$$Z_X = 50 \ \frac{10}{3}$$
$$= 50 \times 3.33$$
$$= 166.5\Omega$$

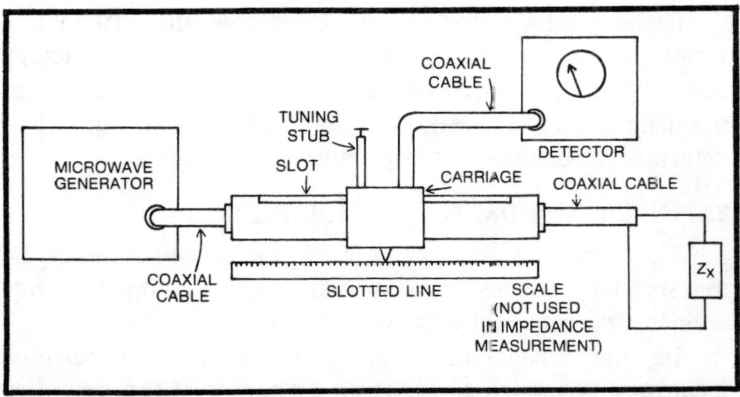

Fig. 3-8. Test setup using the slotted line to measure impedance at microwave frequencies.

Impedance measurement is only one of the uses of a slotted line. This basic microwave tool is also used for determining wavelength, standing-wave ratio (SWR), and insertion loss. *General Radio's 900-LB* slotted line is usable from 300 MHz to 8.5 GHz and somewhat beyond. *Hewlett-Packard's 817A/B* operates from 1.8 GHz to 18 GHz. Some slotted lines are essentially a section of air-dielectric coaxial line; others are essentially a section of waveguide.

3.13 SWR METHOD

Often, the detector used with a slotted line (see Fig. 3-8) is a direct-reading, standing-wave-ratio (SWR) meter. In this instance, the unknown impedance connected to the line may be calculated from the observed SWR and the characteristic impedance (Z_O) of the line:

$$Z_X = Z_O \times \text{SWR} \qquad (3\text{-}17)$$

Example 3-12. The test setup shown in Fig. 3-8 is operated with a 50Ω slotted line in the same manner as described in Sec. 3.12, and the indicated SWR is 1.15. Calculate the unknown impedance.

From Eq. 3-17:

$$Z_X = 50(1.15)$$
$$= 57.5\Omega$$

Radio amateurs and Citizens Band operators often use a simple bridge-type SWR meter (either homemade or factory built), and the SWR obtained with this instrument at frequencies up to 150 MHz also may be used with Eq. 3-17 to determine an unknown RF impedance.

3.14 INPUT IMPEDANCE OF AMPLIFIER

The input impedance (Z_{IN}) of an amplifier may be measured by means of a special calibrated input-voltage divider. The test setup is shown in Fig. 3-9.

In this arrangement, the test signal of a selected frequency is applied to the input terminals of the amplifier through a calibrated variable resistor, R_S. This puts R_S in series with the input impedance (Z_{IN}) of the amplifier. An electronic AC voltmeter/millivoltmeter (VTVM or TVM) is arranged with a switch so that the signal voltage E_1 to the input of R_S and Z_{IN} in series may be read when switch S is at position A, and the signal voltage E_2 at the amplifier input terminals may be read when S is at B. The amplifier is switched ON and is terminated by load resistor R_L, whose resistance equals the rated output impedance of the amplifier. The generator output voltage E_1 must be chosen such that the amplifier is not overdriven during the test. This impedance measurement is based upon the fact that when resistance R_S equals amplifier input impedance Z_{IN}, amplifier input voltage E_2 is half of the generator output voltage E_1. It is necessary only to adjust resistance R_S to the point at which E_2 (switch S at position B) equals 0.5 E_1 (switch S at position A), whereupon the value of amplifier input impedance may be read from the ohms-calibrated dials of R_S.

Test Procedure

- Set up test circuit as shown in Fig. 3-9.
- Load resistance R_L must be equal to the rated output impedance of the amplifier and must be capable of handling at least twice the rated output power of the amplifier.
- With variable resistor R_S set to maximum resistance, throw switch S to position A and adjust output of the generator to a voltage level that will not overload the amplifier. Record voltage as E_1.
- Throw switch S to position B and readjust R_S until meter reads half of E_1. Record as E_2.
- Return S to position A and recheck E_1. If E_1 is not exactly twice E_2, reset resistor R_S, return S to B, and recheck E_2.
- Continue to throw S back and forth between A and B while checking the voltage at each position of the switch.
- When E_2 is exactly 0.5 E_1, and E_1 remains exactly $2E_2$, the resitance setting of variable resistor R_S equals the

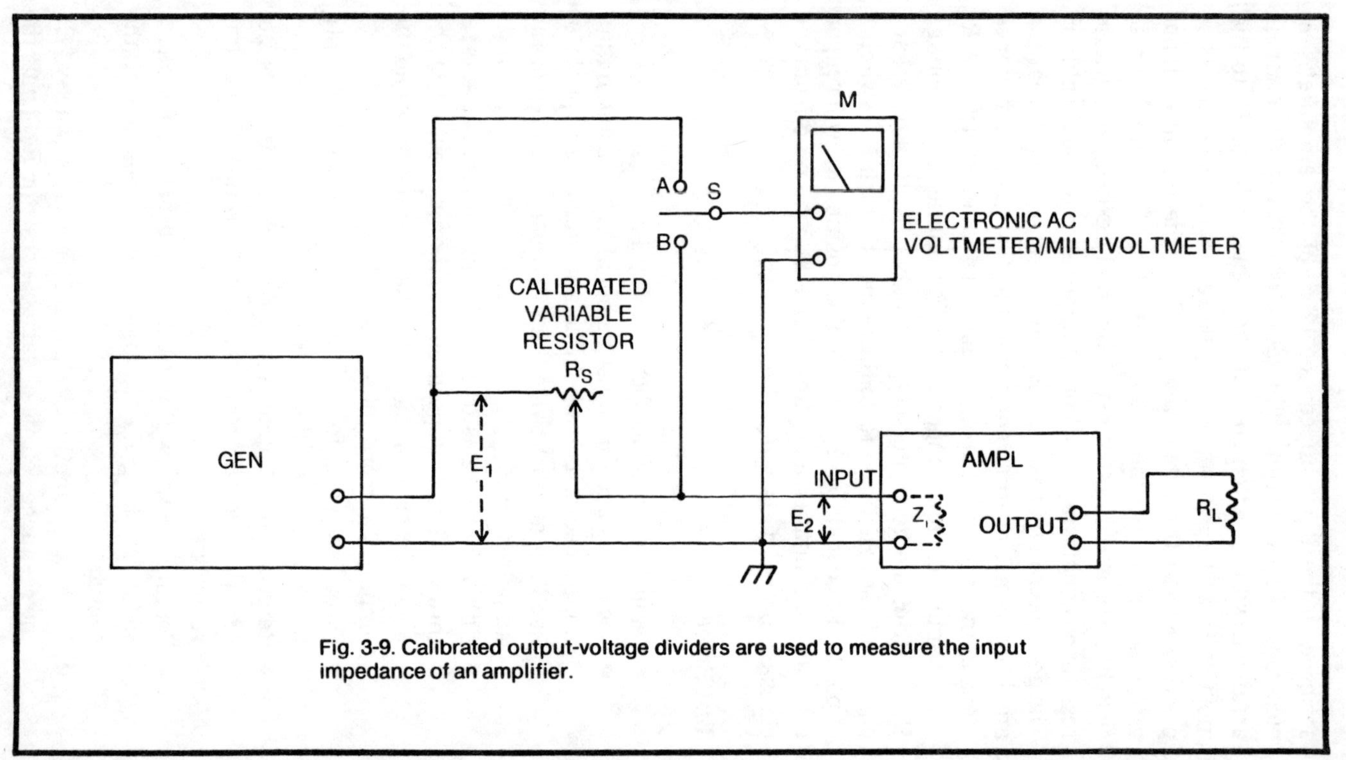

Fig. 3-9. Calibrated output-voltage dividers are used to measure the input impedance of an amplifier.

input impedance of the amplifier: $Z_{IN} = R_S$. If the R_S dial has previously been calibrated in ohms, the impedance may be read directly from the dial; otherwise, R_S must be temporarily disconnected and its resistance setting checked with an ohmmeter or bridge.

The input impedance of an amplifier can be measured also by the *resistance/balance method* (see Sec. 3.6 and Fig. 3-4) if the amplifier input is placed at Z_X in Fig. 3-4. Also, the *voltmeter/ammeter method* (Sec. 3.2 and Fig. 3-1) and the *ammeter method* (Sec. 3.3 and Fig. 3-2) can be used, provided the meters are sensitive enough to indicate the low signal levels required (millivolts or microvolts, and milliamperes or microamperes). In each of these alternate methods of measurement, the amplifier must be switched on.

3.15 OUTPUT IMPEDANCE OF AMPLIFIER

Figure 3-10 shows two methods of measuring the output impedance Z_{OUT} of an amplifier. Each employs a variable load resistor R_L, but Fig. 3-10(a) employs an electronic AC voltmeter (VTVM or TVM) whereas 3-10(b) requires an AC wattmeter. The method selected will depend, in most cases, upon which instrument is immediately available. In each instance the amplifier is driven by a signal of desired frequency supplied by the generator, and the amplitude of this signal is sufficient to drive the amplifier to full output without overloading. The amplifier controls are set for maximum output. In each instance load resistor R_L must be capable of handling at least twice the rated output power of the amplifier.

Resistor/Voltmeter Method

In the arrangement of Fig. 3-10(a), voltmeter *M* is operated by the output signal of the amplifier. When switch *S* is open this meter indicates the no-load output voltage E_1. When *S* is closed the meter indicates the voltage E_2 for full loading of the amplifier by resistor R_L. When R_L is adjusted to the resistance equal to the amplifier output impedance (Z_{ON}), E_2 is $0.5E_1$, since under these circumstances a 2:1 voltage divider is formed by Z_{OUT} and R_L in series (see Secs. 2.7 and 2.8, Ch. 2).

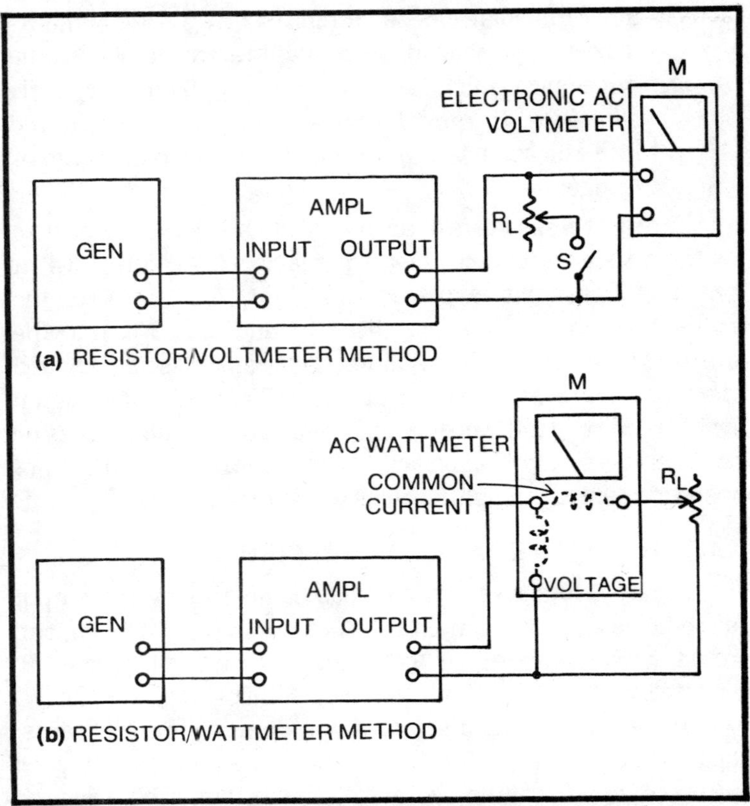

Fig. 3-10. Methods of measuring the output impedance of an amplifier: (a) resistor/voltmeter method; (b) resistor/wattmeter method.

The test procedure consists simply of adjusting R_L to the point at which $E_2 = 0.5E_1$ and reading the output impedance from the ohms-calibrated dial of the resistor. If the resistor has not been calibrated, it may be temporarily disconnected and its resistance setting checked with an ohmmeter or bridge.

Resistor/Wattmeter Method

In the arrangement of Fig. 3-10(b) wattmeter *M* indicates the amplifier output power in variable load resistor R_L. When the resistance of R_L equals the output impedance Z_{OUT} of the amplifier, the power in the load is maximum (see Sec. 2.10, Ch. 2). The resistor must be rated to handle at least twice the expected output power of the amplifier (heavy duty rheostat/potentiometers are available for this purpose).

The test procedure consists simply of adjusting R_L for peak deflection of the wattmeter. At that point, the resistance setting of R_L equals the output impedance of the amplifier, and Z_{OUT} can be read directly from the ohms-calibrated dial of the resistor. If the resistor has not been calibrated, it may be temporarily disconnected and its resistance setting checked with an ohmmeter or bridge.

The reader must be forewarned that the wattmeter response in this test is not sharp, so care must be taken in adjusting resistor R_L near the peak deflection. In this connection, Fig. 3-11 shows the curve for a 20-watt amplifier having an output impedance of 50Ω.

3.16 INPUT AND OUTPUT IMPEDANCE OF RECEIVER

The input impedance (Z_{IN}) of a radio receiver may be measured by the *input voltage divider method* (Sec. 3.14 and

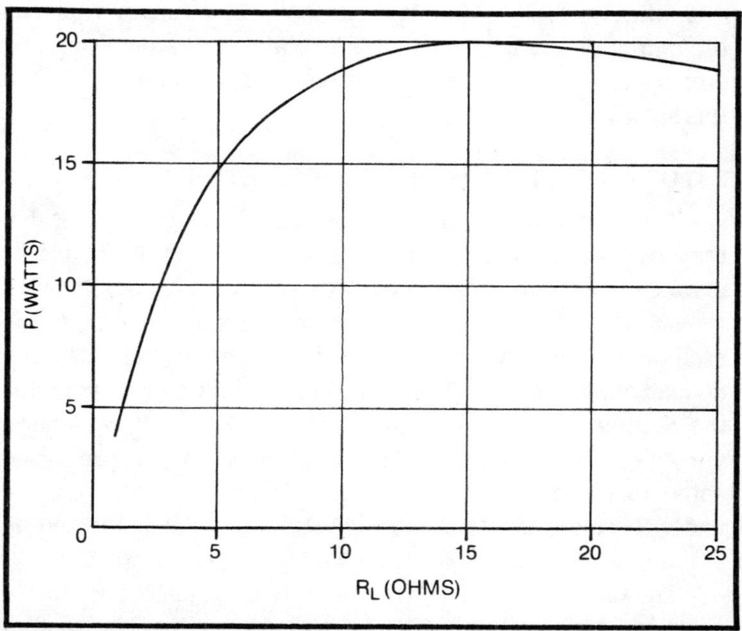

Fig. 3-11. The typical response curve for the resistor/wattmeter method of measuring amplifier output impedance.

Fig. 3-9). The receiver is substituted for the amplifier in Fig. 3-9, variable resistor R_S must be one that will operate satisfactorily at the RF test frequency, and load resistance R_L equals the output impedance of the audio channel. The receiver must be switched on and tuned to the test signals.

The input impedance may be measured also by the *resistance/balance method* (Sec. 3.6 and Fig. 3-4), provided a suitable variable resistor R may be found for the selected RF test frequency. The input terminals of the receiver replace Z_X in Fig. 3-4. The input impedance of receivers is often measured with an RF bridge (see Sec. 3.9 and Fig. 3-6). The *voltmeter/ammeter method* (Sec. 3.2 and Fig. 3-1) and the *ammeter method* (Sec. 3.3 and Fig. 3-2) can be used, provided the meters are sensitive enough to indicate the low signal levels (millivolts or microvolts and milliamperes or microamperes) required. In each of these alternate tests, the receiver should be switched on.

The output impedance of a radio receiver may be measured by the *resistor/voltmeter method* described in Sec. 3.15 and Fig. 3-10(a). For this purpose, the receiver replaces the amplifier in Fig. 3-10(a). A modulated test signal must be employed and the receiver must be switched on and tuned to this signal.

3.17 OUTPUT IMPEDANCE OF OSCILLATOR

The output impedance of an oscillator or signal generator may be measured with the setup shown in Fig. 3-12. In this arrangement, the electronic AC voltmeter/millivoltmeter (VTVM or TVM) is operated by the output signal of the oscillator. When switch S is open this meter indicates the no-load output voltage E_1. When S is closed the meter indicates the voltage E_2 when the amplifier is loaded by calibrated variable resistor R_L. When R_L is adjusted to the resistance equal to the amplifier output impedance E_2 is $0.5E_1$, since under these circumstances a 2:1 voltage divider is formed by Z_{OUT} and R_L in series (see Secs. 2.7 and 2.8, Ch. 2).

The test procedure consists simply of adjusting R_L to the point at which $E_2 = 0.5E_1$, and then reading the output impedance from the ohms-calibrated dial of the resistor. If the resistor has not been calibrated, it may be disconnected temporarily and its resistance setting checked with an ohmmeter or bridge. The resistor must be able to handle safely at least twice the maximum rated output power of the oscillator.

This method may be employed with equal success to check the output impedance of transmitters, industrial oscillators, diathermy machines, and similar equipment.

When an RF generator is under test, resistor R_L and meter M both must be capable of operating at the selected test frequency. Also, the operator must observe closely all of the special precautions common to RF measurements. (See items 5, 10, 11, 12, and 13 in Sec. 3.1.)

3.18 IMPEDANCE OF MECHANICAL GENERATOR

The impedance Z_G of a mechanical generator (rotating machine) may be measured with the setup shown in Fig. 3-13. In this arrangement, as in the one for oscillator measurements described in Sec. 3.17, the electronic AC voltmeter (VTVM or TVM) is operated by the output voltage of the generator. When switch S is open this meter indicates the no-load output voltage E_1. When S is closed the meter indicates the E_2 load voltage (when the generator is loaded by calibrated variable resistor R_L). When R_L is adjusted to the resistance equal to the

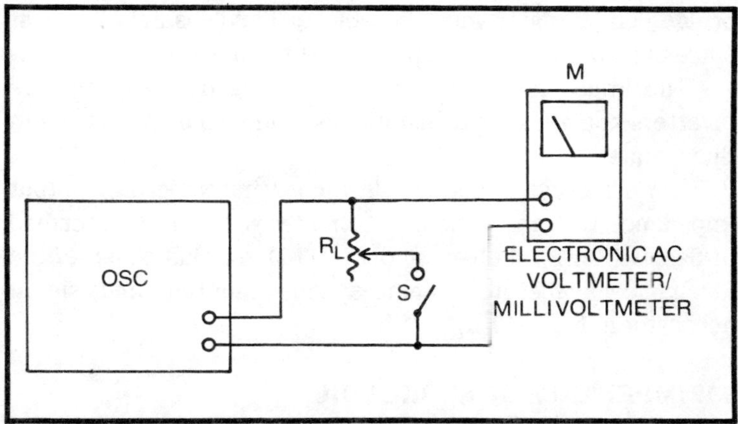

Fig. 3-12. Test setup used to measure the output impedance of an oscillator.

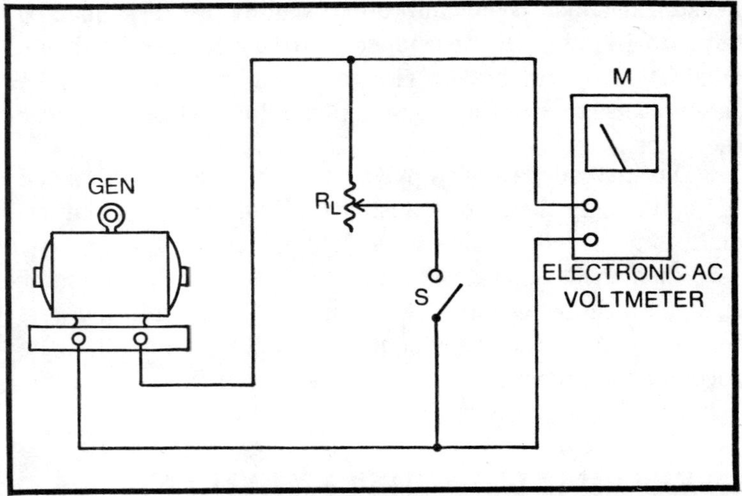

Fig. 3-13. Test setup used to measure the impedance of a mechanical generator.

generator impedance, E_2 is $0.5E_1$, since under these circumstances a 2:1 voltage divider is formed by Z_G and R_L in series (see Secs. 2.7 and 2.8, Ch. 2).

The test procedure consists simply of adjusting R_L to the point at which $E_2 = 0.5E_1$ and reading the output impedance from the ohms-calibrated dial of the resistor. If the resistor has not been calibrated, it may be disconnected temporarily and its resistance setting checked with an ohmmeter or bridge. The resistor must be able to handle safely at least twice the maximum output power of the generator.

The impedance of other AC-producing devices, such as inverters and vibrator transformers, can also be measured in this manner.

Some success is possible in measuring generator output impedance with the *resistor/wattmeter method* as described in Sec. 3.15, Fig. 3-10(b), and Fig. 3-11. In this scheme, the mechanical generator replaces the amplifier and signal generator in Fig. 3-10(b).

3.19 IMPEDANCE OF CHOKE COIL

Iron-core filter chokes are intended to be operated with a specified amount of DC flowing through them. The inductance of such a choke can be widely different under conditions of zero DC and maximum recommended DC, so their impedance should always be measured with the recommended amount of direct current flowing simultaneously with the alternating test current. Any of the following methods may be used to measure choke-coil impedance if means are provided for passing a direct current through the choke during the measurement: Sec. 3.2, Fig. 3-1; Sec. 3.3, Fig. 3-2; Sec. 3.4, Fig. 3-3; Sec. 3.6, Fig. 3-4; and Sec. 3.7, Fig. 3-5. Conventional filtering and bypassing must be added to these circuits to keep the AC test signal out of the DC supply, and vice versa. The AC test-signal amplitude must be no more than 10% of the steady DC level. (See Sec. 1.14, Ch. 1 for a discussion of AC combined with DC.)

For this measurement, some impedance bridges (Sec. 3.8) are equipped with input terminals for a DC component, which is often obtained from an external battery in series with a DC milliammeter and variable resistor. Figure 3-14 shows a typical bridge circuit for choke-coil measurement. This is a *Hay bridge* in which the choke coil L_X and its equivalent series resistance R_X are in one arm; rheostat R_1 in the second arm is the reactance balance; rheostat R_2 in the third arm is the resistance balance; and fixed resistor R_3 in the fourth arm is the ratio resistance which, with capacitor C_2, determines the inductance range of the bridge. Capacitor C_2 is the standard against which the unknown inductance is balanced in the Hay bridge. The variable voltage DC supply is shown here as a battery. The direct current is indicated by DC milliammeter M1 and the AC signal is blocked from the DC circuit by choke L_1. The AC null detector, M2, is an electronic AC voltmeter/millivoltmeter (VTVM or TVM), and the circuit DC is blocked from its input by capacitor C_3 (most such voltmeters have a self-contained input capacitor and do not require C_3).

With B adjusted for the desired direct current through the test choke (L_X), the bridge is separately balanced for reactance (adjustment of R_1) and resistance (adjustment of R_2). At null:

$$L_X = \frac{R_1 R_3 C_2}{1 + (\omega R_2 C_2)^2} \qquad (3\text{-}18)$$

where L_X is the inductance of the choke in henrys, R_1, R_2, and R_3 are in ohms, C_2 is in farads, and $\omega = 2\pi f$, where f is the test frequency in hertz. And:

$$R_X = \frac{R_1 R_2 R_3 (\omega C_2)^2}{1 + (\omega R_2 C_2)^2} \qquad (3\text{-}19)$$

where R, C_2, and ω are in same units as in Eq. 3-18.

Finally, the impedance is calculated from the inductance and resistance values:

$$Z_X = \sqrt{R^2 + (\omega L)^2} \qquad (3\text{-}20)$$

where R and Z_X are in ohms, L is in henrys, and $\omega = 2\pi f$, with f representing the test frequency in hertz.

Example 3-13. A certain power-supply filter choke is checked at 120 Hz in the bridge circuit shown in Fig. 3-14. In this circuit, R_3 is 4000Ω and C_2 is 1 μF. The DC is set to 50 mA. At null, R_1 is set to 5000Ω and R_2 to 90Ω. Calculate the inductance, equivalent series resistance, and impedance of this choke.

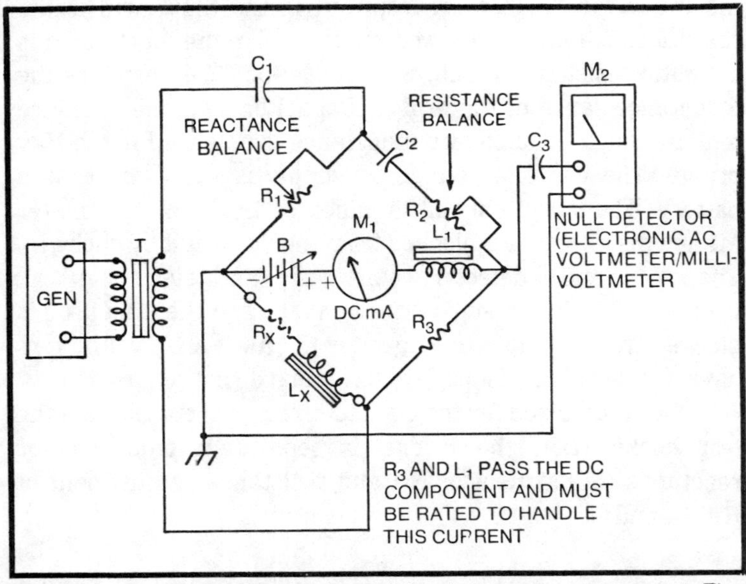

Fig. 3-14. A typical bridge circuit used for choke-coil measurement. The circuit shown is a Hay bridge.

Here, $f = 120$ Hz, $\omega = 754$, and $C_2 = 0.000001$F. From Eq. 3-18:

$$L_X = \frac{5000(4000)0.000001}{1 + (754 \times 90 \times 0.000001)^2}$$

$$= \frac{20}{1 + (0.06786)^2}$$

$$= 20/1.0046$$
$$= 19.9\text{H}$$

From Eq. 3-19:

$$R_X = \frac{5000(90)4000(754 \times 0.000001)^2}{1 + (754 \times 90 \times 0.000001)^2}$$

$$= \frac{1{,}800{,}000{,}000 \times 0.000754^2}{1 + 0.06786^2}$$

$$= \frac{1{,}800{,}000{,}000 \times 0.0000005685}{1 + 0.0046}$$

$$= 1023/1.0046$$
$$= 1018.6\Omega$$

Note: The manfacturer's rating of this choke is 20H, 900Ω.

From Eq. 3-20:

$$Z_X = \sqrt{1018.6^2 + (754 \times 19.9)^2}$$
$$= \sqrt{1{,}037{,}546 + 15{,}005^2}$$
$$= \sqrt{1{,}037{,}546 + 225{,}150{,}025}$$
$$= \sqrt{226{,}187{,}571}$$
$$= 15{,}039\Omega$$

3.20 IMPEDANCE OF CAPACITOR

The AF or RF impedance of a capacitor may be measured by means of any of the following methods described earlier in this chapter, provided the frequency response of the instruments and components is adequate and that the usual precautions are taken at high frequencies: *voltmeter/*

ammeter method, ammeter method, voltmeter method, resistance/balance method, substitution method, impedance bridge, RF *bridge* and *Q-meter method.*

Unless a test frequency is specified, measure AF impedance at 1000 Hz and RF impedance at 1 MHz and 10 MHz.

3.21 IMPEDANCE OF LOUDSPEAKER

The impedance of the voice coil of a loudspeaker may be measured by means of any of the following methods described earlier in this chapter: *voltmeter/ammeter method, ammeter method, voltmeter method, resistance/balance method,* and the *substitution method.*

Whichever method is employed, the test-signal voltage must be kept low in order to minimize the sound emitted by the loudspeaker (when quiet is demanded) and to hold voice-coil current to a safe minimum. It should be noted that the impedance and the DC resistance of a voice coil both are low. Most voice coils have 8Ω impedance; however, common values encountered are 3.2Ω, 3.4Ω, 4Ω, 8Ω, and 16Ω.

The loudspeaker under test should be mounted in the clear so that its cone is not covered but operates in free air. If the loudspeaker normally is operated in a cabinet or behind a baffle, however, it should be so mounted for the test, but again with its front unencumbered.

Unless some other test frequency is specified, impedance will usually be measured at 1000 Hz. For a complete picture of loudspeaker impedance, the unit may be measured at closely spaced frequencies throughout the AF spectrum.

3.22 IMPEDANCE OF HEADPHONES

The impedance of headphones and earplugs is measured in the same way as that of loudspeakers.

Headphones, unlike loudspeakers, are available in several different types over a wide impedance range. Communications-type magnetic headphones, for example, are usually specified as 2500Ω or 5000Ω. Communications-type crystal headphones can exhibit an impedance of 30K–100K. Stereo headphones exhibit low impedance, common values being 3.2Ω, 4Ω, 8Ω, 16Ω, 32Ω, and 600Ω. A small earplug, such as those used with hearing aids and shirtpocket transistor radios, can present a DC resistance of 2K–3K and an AC impedance of 6700Ω.

3.23 IMPEDANCE OF NONLINEAR DEVICES

The small-signal impedance of nonlinear devices is often quite different from their DC resistance at a selected operating point. These devices include conventional semiconductor diodes and rectifiers, zener diodes, tunnel diodes, transistors, lamp filaments, thermistors, voltage-dependent resistors, and saturable reactors.

Figure 3-15 shows the test setup. In this arrangement, the direct current I_{DC} for the desired operating point flows through the nonlinear impedance device Z_X from a variable DC supply shown here as a battery. The value of this current is indicated by DC milliammeter $M1$. Simultaneously, an alternating current I_{AC} is passed through the device; this latter current is

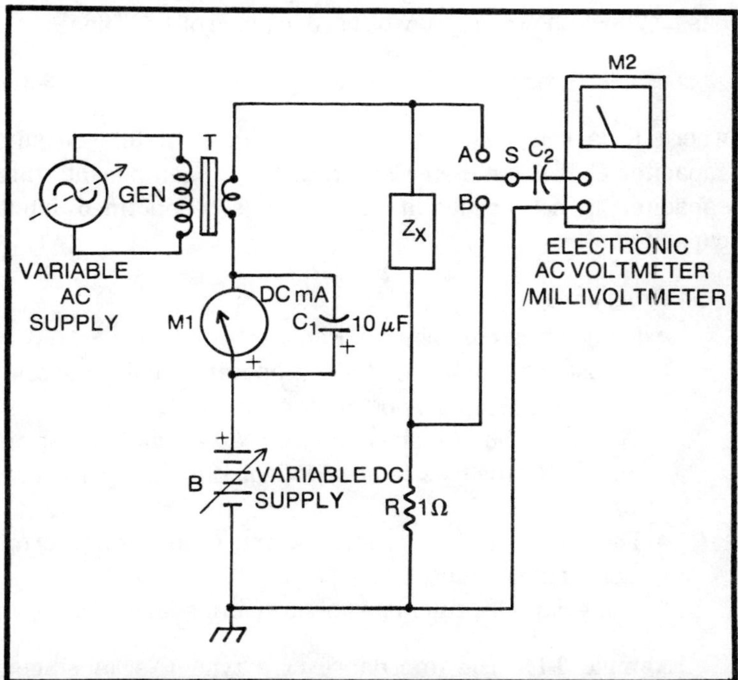

Fig. 3-15. Test setup for measuring the impedance of nonlinear devices.

introduced into the circuit by the low-impedance secondary winding of transformer T and is supplied by a variable AC supply. The RMS value of the current must not exceed one-tenth of the value of the direct current. Bypass capacitor C_1 carries the AC around the DC milliammeter. The AC component produces a voltage drop E_X across resistor R which is proportional to this current, and this voltage is read by the electronic AC voltmeter/millivoltmeter $M2$ (VTVM or TVM) when switch S is in position B. The low resistance of R (1Ω) will in most instances be negligible with respect to Z_X and can be ignored. Meter $M2$ thus becomes a sensitive direct-reading AC milliammeter when S is at B, since $I_{AC} = E_X/R = E_X/1 = E_X$ (I_{AC} is in amps, and E_X is in volts), and milliamperes may be read directly from the voltage scales. When switch S is at position A, meter $M2$ reads the voltage E_Z across the impedance device. From the two readings of this meter, the unknown impedance may be calculated on the basis of $Z_X = E_{AC}/I_{AC}$. Since I_{AC} equals E_X, as has just been shown, E_X may be used in place of I_{AC}. Then,

$$Z_X = E_Z/E_X \qquad (3\text{-}21)$$

where E is RMS volts and Z_X is in ohms. In this circuit, capacitor C_2 isolates meter $M2$ from the DC component; this capacitor is not needed if $M2$ has a self-contained input capacitor.

Test Procedure

- Set up circuit as shown in Fig. 3-15.
- Adjust DC supply for desired operating-point current (I_{DC}) as indicated by DC milliammeter $M1$.
- Throw switch S to position B to read E_X and adjust AC supply for the RMS value of E_X equal to 0.1 I_{DC}. Record as E_X.
- Throw switch S to position A and read voltage drop across impedance. Record as E_Z.
- Using Eq. 3-21, calculate unknown impedance.

Example 3-14. The impedance of a type 1N458A silicon diode is measured at a DC operating point of 100 mA (I_{DC}). The alternating test-signal current must not exceed 0.1 I_{DC}; that is, it must not exceed 10 mA. This corresponds to $E_X = 0.01$V. When switch S is at position B, $E_X = 0.01$V. With S at position A, $E_Z = 0.75$V. Calculate the diode impedance at this 100 mA DC operating point.

From Eq. 3-21:

$$Z_X = E_Z/E_X$$
$$= 0.75/0.01$$
$$= 75\Omega$$

3.24 COMMERCIAL IMPEDANCE INSTRUMENTS

This section briefly describes several commercial instruments for the evaluation of impedance. This equipment is apart from impedance bridges, RF bridges, and Q meters, and the descriptions are arranged alphabetically by name of manufacturer.

Clarke-Hess Model 273 ESR Meter. A digital instrument that automatically indicates *equivalent series resistance* (1 mΩ to 20Ω) of any type of capacitor from 0.005 μF to 1F. The test frequency is 1 MHz. This instrument will also measure the internal resistance of a battery.

General Radio Type 1602-B UHF Admittance Meter. This is a continuously tunable coaxial device which measures complex impedance and admittance. Its frequency range is 40 MHz to 1.5 GHz, and it requires an external generator and external detector.

The tuning dials and scale multipliers of this instrument permit readings directly in conductance G(reciprocal of resistance) from 0.01 to 4000 millimhos, and susceptance β(reciprocal of reactance) from −4000 to +4000 millimhos. From these values, impedance may be calculated: $Z_X = \sqrt{(1/G)^2 + (1/\beta)^2}$. When a constant-impedance quarter-wave line is used with this instrument, the dials read directly in resistance and reactance of the impedance device under test. From these values impedance may be calculated: $Z_X = \sqrt{R^2 + X^2}$. Specified accuracy of the admittance meter for both conductance and susceptance is $\pm3\%$ (plus 0.2 millimho) for zero to 20 millimhos; $\pm3\sqrt{M\%}$(plus 0.2

millimho) above 20 millimhos (where *M* is the scale multiplier), up to 1 GHz; and ±5% (plus 0.2 millimho) to 1.5 GHz.

General Radio Type 1684 Digital Impedance Meter. A digital instrument that separately indicates resistance (1 milliohm to 2 megohms), capacitance (0.1 pF to 200 μF), and inductance (0.1 μH to 200H). Resistance accuracy is ±1% of reading, ±0.05% full scale, ±10 milliohms. Capacitance accuracy is ±1% of reading, ±0.05% full scale, ±1 pF. Inductance accuracy is ±1% of reading, ±0.05% full scale, ±1 μH. Impedance may be calculated from these quantities: $Z_X = \sqrt{R^2 + X^2}$. A self-contained generator supplies a 1000 Hz test signal.

Hewlett-Packard Model 4815A RF *Vector Impedance Meter.* This device is a two-metered RF vector instrument, with one meter that indicates impedance in ohms and another that indicates the phase angle in degrees. The impedance coverage is 1Ω to 100K in nine ranges. The phase coverage is zero to 360 degrees in two ranges.

Impedance accuracy is specified as ±4% of full scale, $\pm(f/30 \text{ MHz} + Z/25\text{K})\%$ of the reading (f = frequency in megahertz, and Z = impedance in ohms).

The self-contained generator is continuously variable from 500 kHz to 108 MHz in five bands. Frequency accuracy is ±2% of the setting.

Industrial Model 1100 Impedance Comparator. This instrument comprises a four-arm bridge in which two arms are precisely matched and the other two arms contain the standard impedance Z_S and the unknown impedance Z_X. When the two impedances match, the bridge is in balance and delivers no output. When, on the other hand, Z_X is lower or higher than Z_S, the bridge becomes unbalanced in proportion to the difference; it delivers a proportionate output signal which is amplified and presented to a phase discriminator. The latter deflects a meter which indicates the percentage by which Z_X differs from Z_S and shows both magnitude and sign.

This instrument is designed for operation at 1000 Hz, 10 kHz, and 100 kHz. It accommodates resistors (3Ω to 10 megohms), capacitors (30 pF to 50 μF), and inductors (10 μH to 100H). Full-scale ranges of the meter are ±0.5%, 2%, 5%, and 20%.

Radiometer Model TRB11 Component Comparator. This instrument affords the direct comparison at 1000 Hz of resistors, capacitors, or inductors with a standard.

Ranges provided are: resistance, 10Ω to 10 megohms; capacitance, 20 pF to 20 μF; and inductance, 1 mH to 10H. When required, a DC polarizing voltage is available, variable from −50V to +20V.

Identification of Manufacturers

Clarke-Hess Communication Research Corp., 43 West 16th St., New York, N.Y. 10011.

General Radio Company. 300 Baker Ave., Concord, Mass. 01742.

Hewlett-Packard Co., 195 Page Mill Rd., Palo Alto, Calif. 94306.

Industrial Test Equipment Co., 21 Yennicock Ave., Port Washington, N.Y. 11050.

Radiometer., The London Company, 811 Sharon Drive, Cleveland, Ohio 44145.

3.25 PRACTICE EXERCISES

3.1. In a voltmeter/ammeter test setup the current is 10 mA and the voltage drop 3.1V. Calculate the unknown impedance in ohms.

3.2. In a voltmeter/ammeter test setup the current is 0.76A and the voltage drop 1.5 mV. Calculate the unknown impedance in milliohms.

3.3. In a voltmeter/ammeter test setup the current is 1A and the voltage drop 0.25V. Calculate the unknown impedance in ohms.

3.4. In a voltmeter/ammeter test setup the current is 500 μA and the voltage drop 1V. Calculate the unknown impedance in ohms.

3.5. In an AC ammeter test setup a constant 1V source is used with a 0-1 milliammeter having an internal resistance of 500Ω. Calculate the unknown impedance in ohms when the current is 0.9 mA.

3.6. In an AC ammeter test setup a constant 1V source is used with a 0–1 ammeter having an internal resistance of 0.213Ω. Calculate the unknown impedance in ohms when the current is 0.5A.

3.7. In an AC ammeter test setup a constant 10V source is used with a 0–10 milliammeter having an internal resistance of 1650Ω. Calculate the unknown impedance in ohms when the current is 5.6 mA.

3.8. In an AC ammeter test setup a constant 6.3V source is used with a 0–50 milliammeter having an internal resistance of 80Ω. Calculate the unknown impedance in ohms when the current is 40 mA.

3.9. In an AC ammeter test setup a constant 12.6V source is used with a 0–100 microammeter having an internal resistance of 3400Ω. Calculate the unknown impedance in kilohms when the current is 75 μA.

3.10. In an AC ammeter test setup a constant 10V source is used with a 0–300 microammeter having an internal resistance of 1800Ω. Calculate the unknown impedance in kilohms when the current is 165 μA.

3.11. In a voltmeter test setup, using a 10Ω standard resistor, the applied voltage is 6.3V and the voltage drop is 1.1V. Calculate the unknown impedance in ohms.

3.12. In a voltmeter test setup using a 5Ω standard resistor, the applied voltage is 10V and the voltage drop is 9.25V. Calculate the unknown impedance in ohms.

3.13. In a voltmeter test setup using a 10Ω standard resistor, the applied voltage is 0.1V and the voltage drop is 33 mV. Claculate the unknown impedance in ohms.

3.14. In a voltmeter test setup using a 25Ω standard resistor, the applied voltage is 150 mV and the voltage drop is 2 mV. Calculate the unknown impedance in ohms.

3.14. In a voltmeter test setup using a 1Ω standard resistor, the applied voltage is 7.5V and the voltage drop is 0.1V. Calculate the unknown impedance in ohms.

3.16. In a voltmeter test setup using a 1Ω standard resistor, the applied voltage is 10V and the voltage drop is 9.8V. Calculate the unknown impedance in ohms.

3.17. In a substitution-type circuit using a comparison resistance of 10Ω, the output voltage is set initially to 99 mV by setting the input voltage to 10V. With the unknown impedance in place, the input voltage must be reset to 1.09V to restore the 99 mV output. Calculate the unknown impedance in ohms.

3.18. With a certain substitution-type circuit the output voltage is 0.25V. The initial input voltage is 4.5V and the final input voltage is 1V. How much higher is the unknown impedance than the comparison (standard) impedance?

3.19. A certain 250 pF capacitor has a Q at 500 kHz of 1500. Calculate the AC resistance in milliohms.

3.20. A certain 0.1 μF capacitor has a Q at 100 kHz of 50. Calculate the AC resistance in ohms.

3.21. A certain 1 mH inductor has a Q at 1 MHz of 125. Calculate the AC resistance in ohms.

3.22. A certain 20H inductor has a Q at 1 kHz of 139.6. Calculate the AC resistance in ohms.

3.23. A certain 50 pF capacitor has a 1000 Hz dissipation factor of 0.0005. Calculate the AC resistance in ohms.

3.24. A certain 8 μF capacitor has a 120 Hz dissipation factor of 0.15. Calculate the AC resistance in ohms.

3.25. With a 300Ω transmission-line setup as in Fig. 3-7(a), the sending-end current is found to be 10 mA and the receiving-end current is 3.3 mA. Calculate the unknown impedance in ohms.

3.26. With a 75Ω transmission-line setup as in Fig. 3-7(a), the sending-end current is found to be 1.5 mA and the receiving-end current is 1.8 mA. Calculate the unknown impedance in ohms.

3.27. A 50Ω slotted line is employed in an RF impedance measurement. The voltage at a maximum point is 10 mV and at the adjacent minimum point is 2.2 mV. Calculate the unknown impedance in ohms.

3.28. A 50Ω slotted line is employed in an RF impedance measurement. The maxima are 0.13V and the minima are 0.09V. Calculate the unknown impedance in ohms.

3.29. An unknown impedance is connected to a 600Ω line and the SWR is found to be 1.05. Calculate the unknown impedance in ohms.

3.30. With an unknown impedance connected to a 300Ω line, what SWR value must be obtained for the unknown impedance to be 800Ω?

3.31. When a transmission line is terminated in its characteristic impedance, what is the resulting SWR value?

3.32. A certain iron-core choke is tested with a 400 Hz Hay bridge (see Fig. 3-14, Ch. 4). At balance $R_1 = 1255\Omega$, $R_2 = 52\Omega$, $R_3 = 1000\Omega$, and $C_2 = 1\ \mu F$. Calculate the choke's (a) inductance L_X in henrys, (b) resistance R_X in ohms, and (c) impedance Z_X in ohms.

3.33. A certain high-current, low-inductance iron-core choke is tested with a 1 kHz Hay bridge (see Fig. 3-14, Ch. 3). At balance $R_1 = 950\Omega$, $R_2 = 32\Omega$, $R_3 = 100\Omega$, and $C_2 = 0.01\ \mu F$. Calculate the choke's (a) inductance L_X in millihenrys, (b) resistance R_X in milliohms, and (c) impedance Z_X in ohms.

(Correct answers are to be found in Appendix D.)

4 Inductance

Inductance is often an important constituent of impedance. For that reason and because many experimenters wind their own inductors and transformers or modify commercial ones, this chapter is included for working information on inductance. Parallel information on capacitance also might be offered, but this would seem superfluous since it is a rare experimenter indeed who would have occasion to build his own capacitor or modify a manufactured one.

4.1 NATURE OF SELF-INDUCTANCE

A current flowing in a coil of wire causes a magnetic field to build up about the coil with energy being stored in this field. After the voltage first is applied, the current builds up (the magnetic field expands) slowly to its maximum value, since the increase is opposed by a *counter emf* which is induced in the coil and has a polarity opposite to that of the applied voltage. When the applied voltage subsequently is removed, the magnetic field collapses into the coil, inducing a current which flows out of the coil in the direction opposite to that of the original current and returns energy to the external circuit.

When the current is alternating, the amount of opposition the coil (*inductor*) offers to the current is directly proportional to the frequency and to a property which is aptly described as *electrical inertia*, since it is this property that causes the coil to oppose any rapid increase or decrease in current. This apparent inertia is called *self-inductance* or just *inductance*. Inductance is measured in henrys (*H*). An inductor has a self-inductance of one henry when a 1V drop is produced across it by a current change of 1A per second. Because the henry is a large unit for some applications, inductance is also measured in millihenrys (thousandths of henrys, abbreviated mH), microhenrys (millionths of henrys, abbreviated μH), and sometimes picohenrys (millionths of microhenrys, abbreviated pH). Table 4-1 shows the relations between these units of inductance.

All electrical conductors possess inductance; however, winding a length of wire into a coil greatly increases the inductance (because this concentrates the magnetic field) so that a desired number of henrys can be obtained in a small space. The inductance of a simple coil depends upon the length and diameter of the coil and the number of turns of wire, as will be shown below. While most inductors are coils of some kind, straight wires also possess inductance; and while this inductance is small—as is explained in Sec. 4.8—it must be taken into account in circuits where even this small amount can generate a significant high-frequency impedance.

4.2 CORELESS SINGLE-LAYER SOLENOID

A common type of inductor, the single-layer solenoid, consists of a coil of wire that is wound with the turns of wire all in one layer and without a magnetic core (Fig. 4-1). Few-turn inductors of this type can be self-supporting, that is, air-wound, as illustrated in Fig. 4-1(a); coils of many turns are wound for mechanical support on a cylindrical dielectric form,

Table 4-1. Conversion Factors for Various Common Inductance Units.

$$\begin{aligned}
\mathrm{H} &= 10^{3}\,\mathrm{mH} = 10^{6}\,\mu\mathrm{H} = 10^{12}\,\mathrm{pH} \\
\mathrm{mH} &= 10^{-3}\,\mathrm{H} = 10^{3}\,\mu\mathrm{H} = 10^{9}\,\mathrm{pH} \\
\mu\mathrm{H} &= 10^{-6}\,\mathrm{H} = 10^{-3}\,\mathrm{mH} = 10^{6}\,\mathrm{pH} \\
\mathrm{pH} &= 10^{-12}\,\mathrm{H} = 10^{-9}\,\mathrm{mH} = 10^{-6}\,\mu\mathrm{H}
\end{aligned}$$

as seen in Fig. 4-1 (b), or are held together by cement. The single-layer construction is suitable for relatively small coils; these units are rated in microhenrys.

For this inductor, the inductance L may be calculated:

$$l = (0.2\,d^2N^2)/3d + 9l) \qquad (4\text{-}1)$$

where d is the diameter of winding in inches, l the length of winding in inches, and N the number of turns.

From this relationship, the required number of turns for a desired inductance is:

$$N = \sqrt{\frac{L\,(3d + 9l)}{0.2\,d^2}} \qquad (4\text{-}2)$$

Example 4-1. A certain single-layer solenoid consists of 115 turns of No. 32 enameled wire close wound on a form 0.75 inch

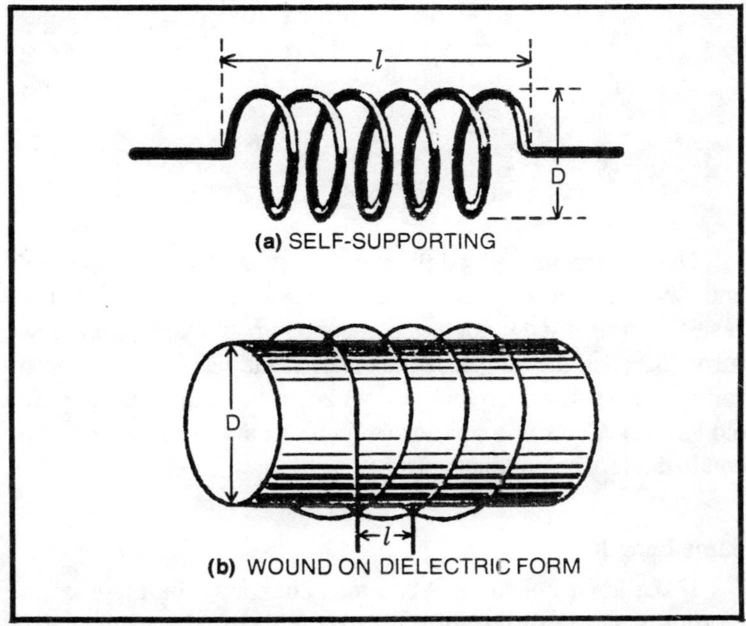

Fig. 4-1. The common single-layer solenoid with turns of wire all on one layer without a magnetic core: (a) self-supporting, (b) wound on a dielectric form.

in diameter. The winding length is one inch. Calculate the inductance in microhenrys.

Here, $N = 115$, $l = 1$, and $d = 0.75$. From Eq. 4-1:

$$L = \frac{0.2 \times 0.75^2 \times 115^2}{(3 \times 0.75) + (9 \times 1)}$$

$$= \frac{0.2 \times 0.5625 \times 13.225}{2.25 + 9}$$

$$= 1487.81/11.25$$
$$= 132.2\ \mu H$$

Example 4-2. A 20 μH single-layer solenoid must be wound in the one-inch winding space of a certain inch-diameter form. How many turns will be required?

Here, $L = 20$, $l = 1$, and $d = 1$. From Eq. 4-2:

$$N = \sqrt{\frac{20((3 \times 1) + (9 \times 1))}{0.2(1^2)}}$$

$$= \sqrt{\frac{20(3 + 9)}{0.2}}$$

$$= \sqrt{240/0.2}$$
$$= \sqrt{1200}$$
$$= 34.6 \text{ turns}$$

When turns are added to an existing single-layer solenoid and the diameter remains unchanged, the value of the resulting increased inductance depends upon whether the new turns increase the length of the original inductor or the length remains the same as before (by squeezing all the turns into the old length). The same applies when turns are removed from a coil to decrease its inductance.

Same Length

If the length of the coil remains constant, the inductance increases as the square of the turns. That is, if the turns are multiplied n times, the inductance increases n^2 times. Conversely, if turns are divided by n, inductance is divided by

$1/n^2$. For example, doubling the number of turns multiplies the inductance by four, the square of two; tripling the number of turns multiplies the inductance by nine, the square of three; etc. Thus, from Eq. 4-1, a 115-turn coil (where $l = 1$ inch and $d = 0.75$ inch) has an inductance of 132.2 μH. If the number of turns is doubled to 230, by using thinner wire in the same winding length, the inductance becomes 529 μH, which is four times the original value. Similarly, if the number of turns is tripled in the same winding length, L becomes 1190.25 μH, which is nine times the original value. Conversely, if the number of turns is halved to 57.5 in the same winding length, L becomes 33.06 μH, one-fourth of the original value; and if the number of turns is reduced one-third to 38.3 in the same winding length, L becomes 14.7 μH, one-ninth of the original value.

Increased or Decreased Length

In a great many cases, changing the number of turns in a coil will alter its length. If the length increases when the number of turns increases, l and N increase while d remains constant. The original inductance then is multiplied by a factor slightly greater than the multiple. For example, consider the 115-turn coil in which $l = 1$ inch, $d = 0.75$ inch, and $L = 132.2$ μH. If the turns are doubled at the same turns-per-inch rate, l becomes 2 inches and L becomes 293.89 μH (2.22 times the original value); and, if the turns are tripled, l becomes 3 inches , and L becomes 457.79 μH (3.46 times the original value). Conversely, if the length of the coil decreases when the number of turns is decreased and the turns are halved to 57.5, l becomes 0.5 inch, and L becomes 55.1 μH (approximately 0.417 times, or slightly less than half the original value); and, if the turns are reduced to one-third of the original, or 38.3, l becomes 0.33 inch, and L becomes 31.34 μH (0.24 times, or somewhat less than one-third the original value). Thus, in this variable-length situation, doubling the turns multiplies inductance by 2+, tripling turns multiplies inductance by 3+, halving turns divides inductance by 2+, and reducing turns to one-third divides inductance by almost 4−.

From these examples, it should be clear that for a constant diameter a larger change in inductance is obtained in a

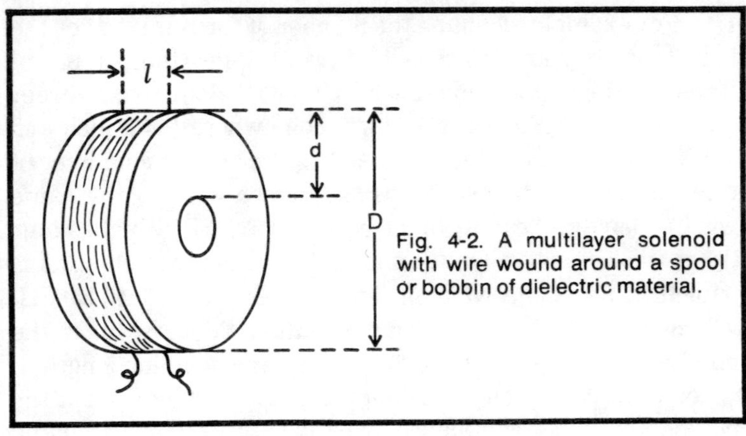

Fig. 4-2. A multilayer solenoid with wire wound around a spool or bobbin of dielectric material.

single-layer coil when added or subtracted turns do not change coil length as compared to when the alteration does change length.

It is a matter of interest that in a single-layer solenoid without magnetic core, the maximum inductance that can be obtained with a given length of wire results when the ratio of the radius to length of the coil is approximately 1.25.

4.3 CORELESS MULTILAYER SOLENOID

High inductance often is obtained by winding a solenoid coil in several layers on a bobbin or spool of dielectric material (Fig. 4-2). The inductance of this coil depends upon length l of the winding, diameter d of the coil, radial depth D_R of the coil, and number of turns N:

$$L = (0.2\,d^2N^2)/(3d + 9l + 10d_R) \qquad (4\text{-}3)$$

where L is the inductance in microhenrys, d the diameter of coil in inches, $D_R =$ the radial depth of coil in inches, l the length of winding in inches, and N the number of turns.

Example 4-3. A certain 1000-turn multilayer solenoid wound on a plastic bobbin has a diameter of 1.25 inches, winding length of 0.75 inch, and radial depth of 0.5 inch. Calculate the inductance in millihenrys.

Here, $N = 1000$, $d = 1.25$, $l = 0.75$, ana $d_R = 0.5$. From Eq. 4-3:

$$L = 0.2 \times 1.25^2 \times 1000^2/_6 (3 \times 1.25) + (9 \times 0.75) + (10 \times 0.5)$$
$$0.2 \times 1.5625 \times 10$$

$$= 312{,}500/15.5$$
$$= 20{,}161\ \mu H$$
$$= 20.16\ mH$$

Because of the complicated cumulative effects of the dimensions of this coil, the inductance changes rapidly with variations in the number of turns. For a given wire size, the number of turns per layer remains the same and so does the winding length, but diameter d and radial depth d_R vary. Thus, if the number of turns given in the foregoing example is halved to 500, diameter d is automatically halved to 0.625 inch, and radial depth D_R to 0.25 inch. From Eq. 4-3, the new inductance then is 1755.62 μH (1.75 mH), approximately 8.7% of the original value.

4.4 COIL WITH STANDARD CORE

The addition of a core of suitable magnetic material (such as iron, powdered iron, ferrite, or nickel alloy) to a coil (Fig. 4-3) increases the coil's inductance, the inductance ideally, if not always so neatly in practice, being multiplied by a number that designates the *permeability* μ of the core material. The permeability of one grade of iron is approximately 2000. Special alloys exhibit very high values; for example, the permeability of Permalloy is as high as 100,000.

The inductance of a coil with magnetic core is given:

$$L = \frac{4.06\, N^2\, \mu\, A}{0.27\, (10^8)\, l)} \tag{4-4}$$

where L is the inductance in henrys, l the total length of core in inches, A the cross-sectional area of core in square inches, N the number of turns, and μ the permeability of core material.

Example 4-4. A 2500-turn coil is wound on a three-inch-long alloy core ($\mu = 5000$) having a cross-sectional area of 0.25 square inch. Calculate the inductance in henrys.

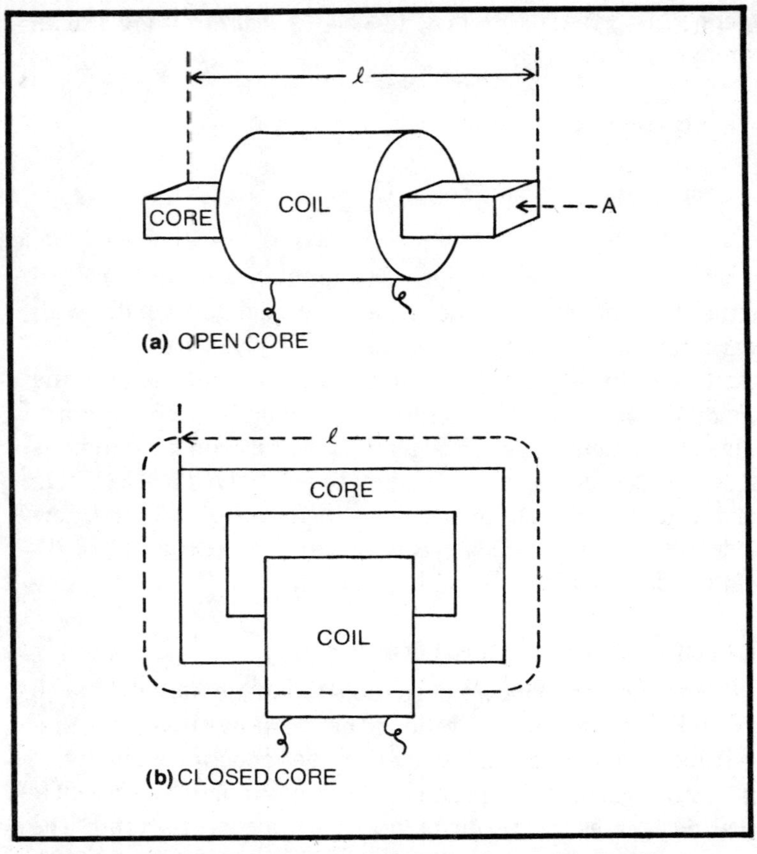

Fig. 4-3. Coils with a magnetic core: (a) open core, (b) closed core.

Here, $N = 2500$, $\mu = 5000$, and $l = 3$. From Eq. 4-4:

$$\begin{aligned} L &= 4.06 \times 2500^2 \times 5000 \times 0.25/1.27 \times 10^8 \times 3 \\ &= 4.06 \times 6{,}250{,}000 \times 5000 \times 0.25/3.81 \times 10^8 \\ &= 3.1718 \times 10^{10} \\ &= 3.81 \times 10^8 \\ &= 83.25\text{H} \end{aligned}$$

For the same core (material, length, and cross section), the inductance varies as the square of the number of turns. That is, if the turns are multiplied n times, the inductance increases n^2 times. Conversely, if the turns are divided by n, the inductance is divided by $1/n^2$. If the number of turns in the foregoing example, for instance, is doubled to 5000, the inductance becomes 333H, four times the original value of 83.25H. And if the turns are halved to 1250, the inductance becomes 20.8H, one-fourth of the original value.

4.5 COIL WITH TOROIDAL CORE

A toroid is an inductor consisting of a coil wound on a toroid (ring- or doughnut-shaped core) of suitable magnetic material. The toroid has the advantages of small size, high Q, compactness, and—above all—self-shielding. Also, if the core is made of ferrite or some other special magnetic material, the inductor can be operated at frequencies of several hundred megahertz. Toroidal construction is illustrated by Fig. 4-4.

The inductance of the toroid is governed by the number of turns in the coil; the permeability of the core material; and the height, inside diameter, and outside diameter of the core:

$$L = 0.011684\, N^2 \mu\, h \log_{10} (OD/ID) \qquad (4\text{-}5)$$

where L is the inductance in microhenrys, N the number of turns, μ the permeability of core material, h the height of the core in inches, OD the outside diameter of the core in inches, and ID the inside diameter of the core in inches.

Example 4-5. 50 turns are wound on a toroid having an outside diameter of 0.75 inch, inside diameter of 0.25 inch,

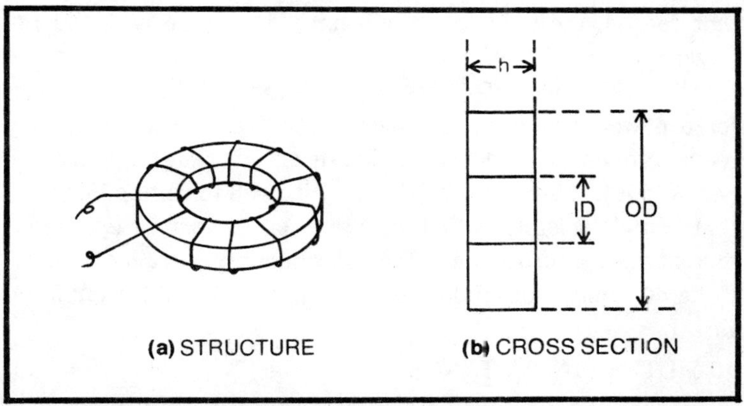

Fig. 4-4. Illustration of coils with toroidal cores showing (a) structure and (b) the cross section.

height of 0.1875 inch, and permeability of 350. Calculate the inductance of this inductor in millihenrys.

Here, $N = 50$, $\mu = 350$, $h = 0.1875$, $OD = 0.75$, and $ID = 0.25$. From Eq. 4-5:

$$\begin{aligned} L &= (0.11684 \times 50^2 \times 350 \times 0.1875) \log_{10}(0.75/0.25) \\ &= (0.11684 \times 2500 \times 350 \times 0.1875) \log_{10} 3 \\ &= 19170(0.47712) \\ &= 9146\ \mu\text{H} \\ &= 9.146\ \text{mH} \end{aligned}$$

Adding or removing turns in a toroid increases or decreases the inductance, respectively, as the square of the number of turns. Thus, if the turns on the same core are multiplied n times, the inductance increases n^2 times. Conversely, if the turns are divided by n, the inductance is divided by $1/n^2$. This means that doubling the turns quadruples the inductance, and removing half the turns reduces the inductance to one-quarter of the original value. If the inductance is to be doubled, 1.41 times as many turns are required (that is, turns must be multiplied by $\sqrt{2}$).

4.6 EFFECT OF DIRECT CURRENT

The inductance of most core-type inductors is affected to some extent by DC flowing through the coil simultaneously with AC. This is because the magnetic properties of the core (especially saturation) are altered temporarily by the DC. For that reason, the inductance will have one value with the DC flowing and another without the DC.

The inductance of a core-type coil that, like a filter choke for a power supply, is intended to carry DC must always be measured with the recommended direct current applied (see Sec. 3.19, Ch. 3 for further details). It is well to remember also that some inductors, which carry no DC in normal use, may exhibit core saturation and the consequent inductance change if the AC signal amplitude is excessive (see Sec. 3.1, item 8, in Ch. 3).

4.7 MUTUAL INDUCTANCE

When the magnetic fields of coils (either separate or wound on the same core) interact, an inductive effect is shared by them. This effect is *mutual inductance* (M). Mutual inductance, like self-inductance, is measured in henrys and in submultiples of the henry (see Table 4-1), and in transformers—where it is of chief interest—is evaluated as follows:

$$M = 4.06\, N_1 N_2\, \mu\, A \qquad (4\text{-}6)$$
$$= 1.27 \times 10^8 \times l$$

where M is the mutual inductance in henrys, N_1 the number of primary turns, N_2 the number of secondary turns, μ the permeability of core material, A the cross-sectional area of core in square inches, and l the total length of core in inches.

Example 4-6. A certain 2:1 transformer is wound on a core having a total length of 8 inches, a cross-sectional area of 1 square inch, and a permeability of 850. The primary coil has 1000 turns and the secondary coil 2000 turns. Calculate the mutual inductance in henrys between the coils.

Here, $N_1 = 1000$, $N_2 = 2000$, $\mu = 850$, $A = 1$, and $l = 8$. From Eq. 4-6:

$$M = 4.06 \times 1000 \times 2000 \times 850 \times 1$$
$$= 1.27 \times 10^8 \times 8$$
$$= 6.902 \times 10^9 = 1.016 \times 10^9$$
$$= 6.79\text{H}$$

4.8 INDUCTANCE OF STRAIGHT, ROUND WIRE

It was mentioned in Sec. 4.1 that even a straight wire possesses inductance. In a very long line this inductance can have a surprisingly significant value. In shorter lengths, straight wires exhibit small inductance, but even this value can be important at very high radio frequencies where a tiny inductance and capacitance can form a resonant circuit.

The inductance of a long, straight, round wire (that is, where the length is at least 1000 times the diameter) is:

$$L = 0.00508\, l\; [\ln(4l/d) - 0.75] \qquad (4\text{-}7)$$

where L is the inductance in microhenrys, l the length in inches, d the diameter in inches, and ln the natural logarithm.

Example 4-7. Calculate the inductance of a straight 10 in. length of No. 24 wire (wire tables give the diameter as 20.1 mils, that is, 0.0201 inch).

Here, $l = 10$, and $d = 0.0201$. From Eq. 4-7:

$$\begin{aligned} L &= 0.00508\,(10)\,[ln\,(4 \times 10)/0.0201 - 0.75] \\ &= 0.00508\,(10)\,(ln40/0.00201 - 0.75) \\ &= 0.0508\,[(ln\,1990) - 0.75] \\ &= 0.0508\,(7.596 - 0.75) \\ &= 0.0508\,(6.846) \\ &= 0.348\ \mu\text{H} \end{aligned}$$

This is a small amount of inductance; nevertheless, the reactance of this length of wire is 218.7Ω at 100 MHz, enough to produce a voltage drop of 2.19V if a 100 MHz current of 10 mA flows through this wire. By comparison, a 10 in. length of much thicker No. 12 wire (diameter = 80.81 mils) has an inductance of 0.277 μH and a 100 MHz reactance of 174Ω. From these facts, it can be seen that even when resistance is neglected, the impedance of short leads can be substantial at very high radio frequencies.

4.9 IMPEDANCE OF INDUCTOR

The impedance of an inductor $Z = \sqrt{R^2 + \omega L^2}$ (see Sec. 2.2, Ch. 2). In an air-core coil operated at 1 MHz or lower, the resistance R is entirely the resistance of the wire in the coil. At high radio frequencies, however, the resistance R is the combined in-phase components due to wire resistance, skin effect, and other losses. In core-type inductors, core losses combine with the wire resistance to determine the full value of R.

Figure 4-5(a) gives an equivalent circuit of a core-type inductor, with losses shown as series resistance components. Here, L is the inductance of the coil, R_C the resistance of the wire in the coil, R_E the eddy-current losses in the core, and R_H the hysteresis losses in the core. This can be simplified to Fig. 4-5(b) in which R_{EQ} is the equivalent resistance corresponding to R_C, R_E, and R_H together.

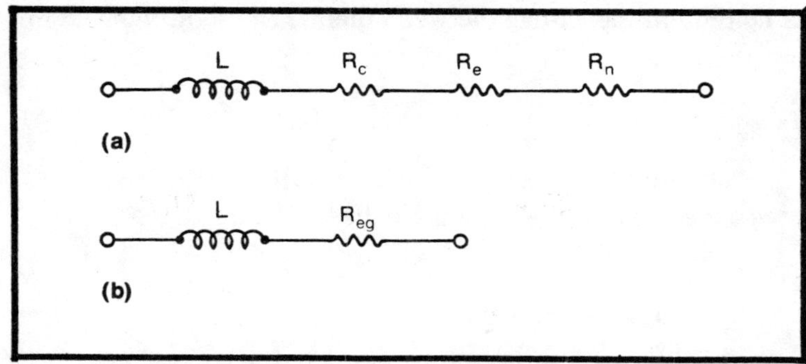

Fig. 4-5. Equivalent circuits of core-type inductors with: (a) losses shown as series-resistant components, and (b) losses simplified to the equivalent resistance.

4.10 BASIC INDUCTOR CIRCUITS

Like resistors and capacitors, inductors may be connected together for lower or higher total inductance. Figure 4-6 shows basic inductor circuits.

When inductors are connected in series, as in Fig. 4-7(a), and positioned so that their fields do not interact (that is, there

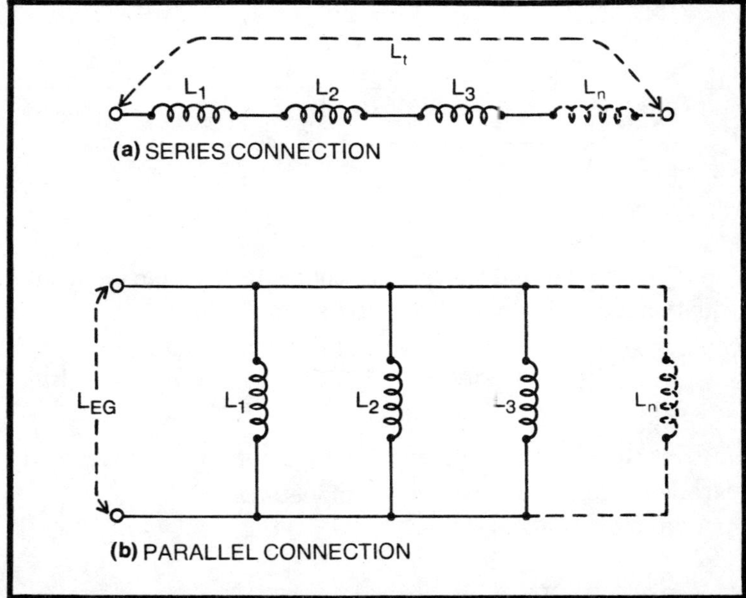

Fig. 4-6. Basic inductor circuits connected in (a) series and (b) parallel.

is no mutual inductance between them), the total inductance is:

$$L_T = L_1 + L_2 + L_3 + ... L_N \qquad (4\text{-}8)$$

Example 4-8. One each of 20H, 5H, 10H, and 18H inductors are connected in series. Calculate the total inductance.

From Eq. 4-8:

$$L_T = 20 + 5 + 10 + 18$$
$$= 53H$$

When inductors are connected in parallel, as in Fig. 4-6(b), and positioned so that their fields do not interact (no mutual inductance between them), the equivalent inductance is:

$$L_{EQ} = \frac{1}{1/L_1 + 1/L_2 + 1/L_3 + ... 1/L_N} \qquad (4\text{-}9)$$

Example 4-9. One each of 20H, 10H, 250 mH, and 1H inductors are connected in parallel. Calculate the equivalent inductance.

From Eq. 4-9:

$$L_{EQ} = \frac{1}{1/20 + 1/10 + 1/0.25 + 1/1}$$

$$= \frac{1}{0.05 + 0.1 + 4 + 1}$$

$$= 1/5.15$$
$$= 0.1942H$$

If only two inductors are connected in parallel, the equation for equivalent inductance is simplified to:

$$L_{EQ} = \frac{L_1 \times L_2}{L_1 + L_2} \qquad (4\text{-}10)$$

Example 4-10. A 10H and 5H inductor are connected in parallel. Calculate the equivalent inductance.

From Eq. 4-10:

$$L_{EQ} = \frac{10 \times 5}{10 + 5}$$

$= 50/15$
$= 3.33H$

4.11 PRACTICE EXERCISES

4.1. Calculate the inductance in microhenrys of a 1 in. diameter single-layer solenoid having 30 turns wound in a space of one inch.

4.2. How many turns will be required for a 100 μH single-layer solenoid having a length of 2 in. and diameter of 1.5 in.?

4.3. The turns of a 50 μH single-layer solenoid are doubled without increasing the length of the coil. What is the final inductance value?

4.4. Sixty turns are removed from a certain 90-turn single-layer solenoid. The final inductance is what percentage of the original inductance?

4.5. Calculate the inductance in millihenrys of a multilayer solenoid having 450 turns, diameter of 1 inch, winding length of 0.5 inch, and radial depth of 0.875 inch.

4.6. Does doubling the number of turns in a multilayer solenoid quadruple the inductance?

4.7. A 350-turn coil is wound on a core having 3.25 inches total length, 0.25 inch cross-sectional area, and a permeability of 800. Calculate the inductance in henrys.

4.8. If the core in exercise 4.7 is replaced with one of the same length and cross-sectional area but with different permeability, what must the new permeability be in order to increase the inductance to approximately 4.82H?

4.9. 100 turns are wound on a toroid having the following specifications: $OD =$ one inch, $ID = 0.625$ inch, $h = 0.1875$ inch, $\mu = 400$. Calculate the inductance in millihenrys.

4.10. If the number of turns is tripled in the coil described in exercise 4.9, what will be the final inductance in microhenrys?

4.11. In a certain 1:1 coupling transformer, the primary and secondary windings have 500 turns each and they are wound on a core having a total length of nine inches, cross-sectional area of 0.39 square inches, and permeability of 2500. Calculate the mutual inductance in henrys.

4.12. The diameter of No. 36 wire is 5 mils (0.005 in.). Calculate the inductance in microhenrys of a straight 2 in. length of this wire.

4.13. Calculate the 50 MHz reactance in ohms of the wire in exercise 4.13.

4.14. One each 50 mH, 1000 μH, and 0.01H inductors are connected in series. Calculate the total inductance in millihenrys.

4.15. One each 10 mH, 1500 μH, and 1H inductors are connected in parallel. Calculate the equivalent inductance in henrys.

(Correct answers are to be found in Appendix D.)

5 General Exercises

5.1. Impedance has the following relation to reactance and resistance: (a) numerical sum of R and X, (b) algebraic sum, (c) vector sum, (d) square of sum.

5.2. When two identical impedances are connected in series the total impedance is the: (a) sum of the two, (b) difference between the two, (c) reciprocal of the two.

5.3. When two identical impedances are connected in parallel the equivalent impedance is (a) higher than either one, (b) lower than either, (c) the square root of the sum of the two.

5.4. An impedance consists of a reactance and resistance; however, a resistance alone is termed an impedance (a) always, (b) only at low frequencies, (c) only below 1000Ω, (d) only at radio frequencies.

5.5. When a resistance and a reactance in series each have the same value the resulting impedance is equal to (a) either the resistance or the reactance, (b) square root of resistance or reactance, (c) sum of resistance and reactance, (d) $\sqrt{2}$ times the resistance or the reactance.

5.6. At low frequencies the resistance component in the impedance of an inductor is (a) resistance of wire in the coil, (b) skin effect, (c) quadrature losses.

5.7. At high radio frequencies the resistance component in the impedance of an inductor is due to (a) resistance of wire in the coil, (b) skin effect, (c) in-phase dielectric losses, (d) resistance of fixtures, (e) all of these.

5.8. The current through an impedance is (a) inversely proportional to the voltage, (b) directly proportional to the voltage, (c) neither of these.

5.9. The voltage drop across an impedance is (a) inversely proportional to the current, (b) directly proportional to the current, (c) neither of these.

5.10. An impedance is (a) inversely proportional to current flowing through it, (b) directly proportional to current, (c) neither of these.

5.11. If the phase angle is known an impedance may be resolved into its R and X components by means of (a) vector diagram, (b) trigonometry, (c) either of these.

5.12. In an impedance device the phase angle between X and R is equal to (a) XR, (b) tan X/R, (c) sin R/X, (d) cos R/X, (e) $\tan^{-1}X/R$, (f) $\sin^{-1}X/R$.

5.13. For an inductive reactance, doubling the frequency (a) doubles the reactance, (b) halves the reactance, (c) multiplies reactance by $\sqrt{2}$, (d) multiplies reactance by 1.5.

5.14. For a capacitive reactance, doubling the frequency (a) doubles the reactance, (b) halves the reactance, (c) squares the reactance, (d) divides the reactance by $\sqrt{2}$.

5.15. In a series RC circuit in which $X_C = R$, the phase angle in degrees is (a) 30, (b) 15, (c) 20, (d) 75, (e) 45.

5.16. In a resonant circuit containing X_L and X_C, X_L equals (a) zero, (b) twice X_C, (c) X_C, (d) $0.5X_C$, (e) none of these.

5.17. In a circuit containing X_L and X_C the combined reactance is (a) $X_L - X_C$, (b) $X_L + X_C$, (c) neither of these.

5.18. The impedance at the center of a horizontal half-wave antenna high above the ground is approximately (a) 48Ω, (b) 73Ω, (c) 600Ω, (d) 75Ω.

5.19. When a generator drives a load the load impedance is (a) in series with the generator impedance, (b) in parallel with the generator impedance, (c) both of these.

5.20. When a load impedance Z_L is transformer coupled the reflected impedance is equal to Z_L times (a) turns ratio, (b) 0.5

turns ratio, (c) square of turns ratio, (d) square root of turns ratio.

5.21. When a generator (impedance = Z_G) drives a load (impedance = Z_L) maximum power is transferred when (a) $Z_L > Z_G$, (b) $Z_L = 10Z_G$, (c) $Z_L = 0.5Z_G$, (d) $Z_L < Z_G$.

5.22. The impedance ratio of a transformer equals (a) turns ratio, (b) square root of turns ratio, (c) 0.5 turns ratio, (d) square of turns ratio.

5.23. The turns ratio of a transformer equals (a) impedance ratio, (b) square root of impedance ratio, (c) 0.5 impedance ratio, (d) square of impedance ratio.

5.24. The input impedance of a quarter-wave line is (a) directly proportional to the square of the characteristic impedance of the line, (b) inversely proportional to the square of the characteristic impedance, (c) neither of these.

5.25. Where Z_{IN} and Z_{OUT} are the input and output impedances of a quarter-wave line the characteristic impedance of the line equals (a) $Z_{IN}Z_{OUT}$, (b) Z_{IN}/Z_{OUT}, (c) $Z_{IN}Z_{OUT}^2$, (d) $Z_{OUT} - Z_{IN}$ (e) $\sqrt{Z_{IN}Z_{OUT}}$.

5.26. The characteristic impedance of a solid-dielectric line is (a) higher than that of an open-air line, (b) lower than that of an open-air line, (c) equal to that of an open-air line.

5.27. In an impedance-matching cathode follower the output impedance is (a) lower than the cathode resistance, (b) higher than the cathode resistance, (c) equal to the cathode resistance.

5.28. In an impedance-matching FET source follower the output impedance is (a) lower than the source resistance, (b) higher than the source resistance, (c) equal to the source resistance.

5.29. Attenuation in nepers is (a) lower than decibels, (b) higher than decibels, (c) equal to decibels.

5.30. Power factor equals (a) $\sin \theta$, (b) $\tan \theta$, (c) $\cos \theta$, (d) none of these.

5.31. Power factor equals (a) Z/R, (b) R/Z, (c) neither of these.

5.32. Figure of merit or Q equals (a) R/X, (b) $\sqrt{R/X}$, (c) R/X^2, (d) X/R, (e) none of these.

5.33. The voltage drop across an AC ammeter equals (a) ER_M, (b) E/R_M, (c) $\sqrt{ER_M}$, (d) $E-R_M$.

5.34. In the resistance/balance method of impedance measurement the ratio of resistance voltage drop to impedance voltage drop is (a) 2, (b) 0.5, (c) 0.1, (d) 4, (e) 1.

5.35. In the standing-wave method of measuring impedance, the unknown impedance is found by multiplying the SWR by the (a) characteristic impedance of the line, (b) input impedance of the line, (c) output impedance of the line, (d) product of input and output impedance of the line.

5.36. When the load impedance equals the generator impedance the load voltage is equal to (a) the open-circuit voltage of the generator, (b) one-tenth of the open-circuit voltage of the generator, (c) half the open-circuit voltage of the generator, (d) none of these.

5.37. The reciprocal of impedance is called (a) susceptance, (b) reluctance, (c) conductance, (d) admittance, (e) remittance.

5.38. The reciprocal of resistance is called (a) susceptance, (b) reluctance, (c) conductance, (d) admittance, (e) remittance.

5.39. The reciprocal of reactance is called (a) susceptance, (b) conductance, (c) reluctivity, (d) admittivity, (e) none of these.

5.40. Impedance of one gigohm is (a) one million ohms, (b) ten million ohms, (c) one thousand megohms, (d) ten billion ohms.

5.41. Impedance of one teraohm is (a) $10^{12}\Omega$, (b) $10^{11}\Omega$, (c) $10^{18}\Omega$, (d) $10^{20}\Omega$.

5.42. Impedance of one megohm is (a) 10,000K, (b) 1000K, (c) 100K, (d) none of these.

5.43. Impedance of one milliohm is (a) 0.10 megohm, (b) 0.1 kilohm, (c) 0.100 ohm, (d) 0.01Ω (e) 0.001Ω.

5.44. Impedance of one kilohm is (a) 0.1 megohm, (b) 0.2 megohm, (c) 0.01 megohm, (d) 0.001 megohm, (e) none of these.

5.45. *Period* is the time duration of one cycle?

5.46. Numerically, is *period* $0.1/f$?

5.47. In a sine-wave cycle, $\theta = 2\pi ft$?

5.48. Does angular velocity equal 6.28 times the frequency?

5.49. Does one radian equal 58 degrees?

5.50. In a square wave are maximum value, RMS value, and instantaneous value equal?

5.51. In an impedance measurement made with a distorted wave can the error be as high as the percentage of distortion?

5.52. Is phase angle equal to the tangent of the reactance-to-resistance ratio?

5.53. Does pure resistance introduce only a 1° phase shift?

5.54. Does pure reactance introduce a 90° phase shift?

5.55. At the same frequency, are inductive reactance and capacitive reactance always equal?

5.56. Does capacitive reactance introduce lagging phase shift?

5.57. In a resonant circuit containing R, C, and L, does the resistance (R) disappear?

5.58. As frequency decreases, inductive reactance decreases and capacitive reactance increases?

5.59. When inductive reactance and capacitive reactance are both present in a circuit, the combined reactance is equal to $X_L - X_C$?

5.60. Impedance is the sum of the squares of resistance and reactance?

5.61. Do series-impedance circuits and parallel-impedance circuits have different equations for total impedance?

5.62. The horizontal (resistance) component of an impedance is equal to $Z \sin \theta$?

5.63. Are standing waves, usable for impedance measurement, present on a transmission line that is terminated in its characteristic impedance?

5.64. Maximum power is transferred when a generator is terminated in its internal impedance?

5.65. In most cases, the output impedance of a device appears as a series impedance?

5.66. In an impedance-matching transformer the turns ratio equals the square of the impedance ratio?

5.67. The length of an impedance-measuring transmission line can be any even multiple of a quarter-wavelength?

5.68. A suitable section of transmission line is usable as a linear transformer for RF impedance matching?

5.69. Would an advantage of an active follower (tube or transistor) be its ability to provide power gain?

5.70. A resistor-pad attenuator matches impedance while providing a desired amount of attenuation?

5.71. In conjugate impedance, X has a value common with that in another impedance, but a different resistance value is encountered. Is this statement *true* or *false*?

5.72. Two impedances, Z_1 and Z_2, are reciprocal when $Z_1Z_3 = Z_2$?

5.73. Frequencies at which the driving-point imvedance of a two-terminal reactive network is zero are termed *poles of impedance*?

5.74. Is *power factor* equal to the cosine of the phase angle?

5.75. Figure of merit: or Q is the ratio of reactance to resistance?

5.76. The ammeter method is convenient for measuring impedance, since the internal resistance of the meter has no effect?

5.77. Are transmission-line methods of measuring RF impedance limited to those high frequencies at which the physical length of the line is not prohibitive?

5.78. Whereas the slotted line is a recognized tool for microwave measurements, simpler SWR meters are unacceptable for RF impedance measurements. Is this statement *true* or *false*?

5.79. When an amplifier drives a load impedance equal to the output impedance of the amplifier, is the load voltage one-half of the open-circuit (no-load) output voltage of the amplifier?

5.80. The impedance of an iron-core filter choke should be measured with the rated direct current flowing through the choke?

5.81. True impedance is always a frequency-dependent property?

5.82. Impedance varies more rapidly with change of the resistive component than with change of the reactive component?

(Correct answers are to be found in Appendix D.)

Appendixes

Appendix A

Impedance Conversion Factors

TO CONVERT		MULTIPLY BY
FROM	TO	
gigohms	kilohms	10^{6}
gigohms	megohms	10^{3}
gigohms	microhms	10^{15}
gigohms	milliohms	10^{12}
gigohms	ohms	10^{9}
gigohms	teraohms	10^{-3}
kilohms	gigohms	10^{-6}
kilohms	megohms	10^{-3}
kilohms	microhms	10^{9}
kilohms	milliohms	10^{6}
kilohms	ohms	10^{3}
kilohms	teraohms	10^{-9}
megohms	gigohms	10^{-3}
megohms	kilohms	10^{3}
megohms	microhms	10^{12}
megohms	milliohms	10^{9}
megohms	ohms	10^{6}
megohms	teraohms	10^{-6}
mecrohms	gigohms	10^{-15}

TO CONVERT FROM	TO	MULTIPLY BY
microhms	kilohms	10^{-9}
microhms	megohms	10^{-12}
microhms	milliohms	10^{-3}
microhms	ohms	10^{-6}
microhms	teraohms	10^{-18}
milliohms	gigohms	10^{-12}
milliohms	kilohms	10^{-6}
milliohms	megohms	10^{-9}
milliohms	microhms	10^{3}
milliohms	ohms	10^{-3}
milliohms	teraohms	10^{-15}
ohms	gigohms	10^{-9}
ohms	kilohms	10^{-3}
ohms	megohms	10^{-6}
ohms	microhms	10^{6}
ohms	milliohms	10^{3}
ohms	teraohms	10^{-12}
teraohms	gigohms	10^{3}
teraohms	kilohms	10^{9}
teraohms	megohms	10^{6}
teraohms	microhms	10^{18}
teraohms	milliohms	10^{15}
teraohms	ohms	10^{12}

Appendix B

Phase Angle Data

ϕ (degrees)	Θ (radians)	X/R
10	0.17453	0.17633
15	0.26180	0.26795
20	0.34906	0.36397
25	0.43633	0.46631
30	0.52360	0.57735
35	0.61086	0.70021
40	0.69813	0.83909
45	0.78540	1.00000
50	0.87266	1.19175
55	0.95993	1.42815
60	1.04720	1.73205
65	1.13446	2.14451
70	1.22173	2.74748
75	1.30899	3.73205
80	1.39626	5.67128
85	1.48353	11.43005
90	1.57080	∞

Appendix C

Abbreviations Used in This Book

A — amperes; cross-sectional area of a coil
AC — alternating current
AF — audio frequency
AMPL — amplifier
arc tan — the angle corresponding to a given tangent; also written tan^{-1}
B — battery
β — reciptocal of reactance
C — capacitance; capacitor
cos — cosine of angle
cosech — hyperbolic cosecant
cosh — hyperbolic cosine
coth — hyperbolic cotangent
D — distortion; dissipation factor; depth
d — diameter; differential of
dB — decibels
DC — direct current
E — voltage
e — instantaneous voltage
E_{AC} — alternating-current voltage
E_{AVG} — average value of AC voltage
E_C — voltage across capacitor
E_G — generator voltage

e_G—grid voltage
E_H—harmonic voltage
E_L—load voltage; voltage across inductor
E_{MAX}—maximum value of voltage
emf—electromotive force
E_{MIN}—minimum value of voltage
E_O—output voltage
E_P—primary voltage
E_P—plate voltage
E_R—voltage across resistor
E_{RMS}, V_{RMS}—effective (root mean square) value of AC voltage
E_S—secondary voltage
e_S—screen voltage
E_T—total (or combined) voltage
E_{TERM}—terminal voltage
E_X—unknown voltage
E_Z—voltage across impedance
F—farad; Fahrenheit
f—frequency; fundamental frequency
f_R—resonant frequency
G—$n \times 10^9$; conductance; reciprocal of resistance
GEN—generator
g_{FS}—forward transconductance of FET
GHz—gigahertz
g_M—transconductance
H—henry
h—harmonic; height of core
h_{FE}—forward-current transfer ratio of bipolar transistor
h_{IE}—input impedance of bipolar transistor
Hz—hertz
I—current
i—instantaneous current
I_{AC}—alternating current
I_{AVG}—average value of alternating current
i_B—base current
i_C—collector current
ID—inside diameter
i_E—emitter current
i_G—grid current

I_{MAX}—maximum value of current
I_{MIN}—minimum value of current
I_P—primary current
i_P—plate current
I_{PEAK}—peak value of alternating current ($I_{PEAK} = 1.414\, I_{RMS}$)
I_R—current in a resistance
I_{RMS}—effective (root mean square) value of alternating current
I_S—secondary current
i_S—screen current
I_Z—current in an impedance
K—ohms × 1000; kilohms
k—dielectric constant; any constant
kHz—kilohertz
l—length of winding
L—inductance; inductor
L_{EQ}—equivalent inductance
ln—natural logarithm
l_N—most remote value of inductance
log—common logarithm, for example, $\log_{10}$
L_T—total inductance
L_X—unknown inductance
M—multiplier; megohms; mutual inductance; meter
mA—milliamperes
mH—millihenrys
MHz—megahertz
ms; msec—milliseconds
mV—millivolts
n—transformer turns ratio; attenuation ratio
n—any remote number
N_P—number of primary turns in a transformer
N_S—number of secondary turns in a transformer
ns; nsec—nanoseconds
OD—outside diameter
P—power
pF—picofarads
pf—power factor
P_L—load power
P_T—total power
Q—figure of merit: $Q = X/R$; quality factor; symbol for transistor

R — resistance; resistor
R_{AC} — alternating-current resistance
rad — radians
r_B — bias resistor
R_C — resistance of a coil
r_E — emitter resistor
R_E — resistive losses due to eddy currents
R_{EQ} — equivalent resistance
RF—radio frequency
R_G — generator resistance; gate resistance of FET
r_G — grid resistor
r_K — cathode resistor
R_L — load resistance
R_M — internal resistance of meter
r_{OSS} — output resistance of FET
r_P — plate resistance
R_R — radiation resistance
R_{REFL}—reflected resistance
r_S — source resistor in FET circuit
R_T — total resistance
R_X — unknown resistance
S — switch
s; sec—seconds
sin — sine of angle
sinh — hyperbolic sine
SWR — standing-wave ratio
T — transformer
t — period; time
tan — tangent of angle
tanh — hyperbolic tangent
TV — television
TVM — transistorized voltmeter
V — volts; velocity factor; symbol for electron tube
VA — voltamperes
v_B — base voltage
v_C — collector voltage
VDR — voltage-dependent resistor
v_E — emitter voltage
VTVM — vacuum-tube voltmeter

W—watts
X—reactance; total reactance; unknown quantity
x—horizontal axis
X_C—capacitive reactance
X_L—inductive reactance
Y—admittance
y—vertical axis
Z—impedance
z_B—base impedance
Z_0; Z_o—characteristic impedance
z_C—collector impedance
z_E—emitter impedance
Z_{EQ}—equivalent impedance
Z_G—generator impedance
z_G—grid impedance
Z_{IN}—input impedance
Z_L—load impedance
Z_M—mutual impedance
Z_N—the remotest impedance in a combination
Z_{OUT}—output impedance
Z_P—primary impedance; plate impedance
Z_{REF}—reflected impedance
Z_S — standard impedance; secondary impedance; source impedance
Z_T—total impedance
Z_M—internal impedance of voltmeter
Z_X—unknown impedance

SYMBOLS

Δ — difference between two successive values of a quantity; change
θ—angle; phase angle (radians); attenuation (in nepers)
ϕ—phase; phase angle (degrees)
λ—wavelength
μ— × 0.000001; amplification factor; permeability; micron
μF—microfarads
μH—microhenrys

μs — microseconds

π—the constant 3.14159+; the value of π given to nine decimal places by a pocket calculator is 3.141592654

Ω — ohms

ω — angular velocity $2\pi f$

$>$ — is greater than

$<$ — is less than

Appendix D

Answers to Practice Exercises

Chapter 1

1.1. 0.2505 MHz
1.2. 10,000 MHz
1.3. 3550 kHz
1.4. 0.06 kHz
1.5. 8×10^{9} Hz
1.6. 0.2 μs
1.7. 16.67 ms
1.8. 6×10^{-7}s
1.9. 0.02s
1.10. 1000 μs
1.11. 0.00025 ms
1.12. 1.85 μs
1.13. 3.68×10^{-8}s
1.14. 9.35×10^{-5} ms
1.15. 0.0175 μs
1.16. 1×10^{-9}s
1.17. 3.33×10^{-6} ms
1.18. 0.000125 μs
1.19. 100 Hz
1.20. 6.67 kHz
1.21. 0.1 MHz
1.22. 10 GHz
1.23. 120 Hz
1.24. 2 kHz
1.25. 1 MHz
1.26. 0.5 GHz
1.27. 1000 Hz
1.28. 14.28 kHz
1.29. 0.1 MHz
1.30. 0.2 GHz
1.31. 114.98V
1.32. −2.95V
1.33. 10V
1.34. Zero volts
1.35. 0.625 μs and 0.875 μs
1.36. 0.208 ms and 1.04 ms
1.37. 10,000 Hz
1.38. −7.07V
1.39. (a) 0.1 μs, (b) 0.3 μs
1.40. 200 Hz
1.41. 0.689 radian
1.42. 0.0916 radian

1.43. 309.4 degrees
1.44. 60 degrees
1.45. 0.785 radian
1.46. 0.6283 radian
1.47. (a) 1.57 radian, (b) 4.71 radians
1.48. 180 degrees
1.49. 270 degrees
1.50. (a) 251.3
(b) 785.4
(c) 5026.4
(d) 628,300
(e) 3,392,820
(f) 8,670,540
(g) 1.178×10^7
(h) 6.723×10^7
(i) 1.702×10^8
(j) 3.393×10^8

1.51. 6283 kHz
1.52. 10.6V
1.53. 1.67 μV
1.54. 5.67V
1.55. 6.37V
1.56. 3.66V
1.57. 0.000167V
1.58. 70.7V
1.59. 1.41 μV
1.60. 459.5V
1.61. 34.24 mV
1.62. (a) 10%
(b) 2.5%
(c) 1%

1.63. 0.104%
1.64. 6.59%
1.65. 2.5 mV
1.66. 2261.9V
1.67. 16.71V
1.68. 11.309K
1.69. 628.3Ω
1.70. 3.183H
1.71. 79.5 Hz
1.72. 628.3V
1.73. 20Ω
1.75. 0.00318H
1.76. 79.58Ω
1.77. 127.3Ω
1.78. 165.8Ω
1.79. 106.1 megohms
1.80. 79.58 Hz
1.81. 6.28 mA
1.82. 47.74V
1.83. 10K
1.84. 0.0397 μF
1.85. 1.59 mV
1.86. 5.77 μF
1.87. 40Ω
1.88. 12.5Ω
1.89. (a) 14.91K
(b) inductive

1.90. (a) 15.95Ω
(b) inductive

1.91. 22.51 kHz
1.92. 0.833 MHz
1.93. 0.0316 μF
1.94. 6.25 pF
1.95. 41.35 mH
1.96. 3.13H

Chapter 2

2.1. 105Ω
2.2. 933.3Ω
2.3. 100 mV
2.4. 6.4V
2.5. 0.4 mA
2.6. 1.77A

2.7. 1.38K
2.8. 2.5×10^{-5} gigohms
2.9. 0.58 megohm
2.10. 5×10^{5} microhms
2.11. 100 milliohms
2.12. 9.35×10^{-7} teraohm
2.13. 0.001 gigohm
2.14. 0.5 megohm
2.15. 1×10^{8} microhms
2.16. 5×10^{4} milliohms
2.17. 33,000Ω
2.18. 5.35×10^{-5} teraohm
2.19. 5.163 gigohms
2.20. 1×10^{6}K
2.21. 1×10^{10} microhms
2.22. 1×10^{6} milliohms
2.23. $4.7 \times 10^{6}\Omega$
2.24. 0.05 teraohm
2.25. 1×10^{6}K
2.26. 250 megohms
2.27. 1×10^{14} microhms
2.28. 1×10^{9} milliohms
2.29. $2 \times 10^{9}\Omega$
2.30. 0.0003 teraohm
2.31. 7000 gigohms
2.32. 1.52×10^{10}K
2.33. 2×10^{7} megohms
2.34. 1×10^{16} microhms
2.35. 1×10^{12} milliohms
2.36. $8 \times 10^{11}\Omega$
2.37. 1×10^{-12} gigohm
2.38. 5.52×10^{-6}K
2.39. 1×10^{-8} megohm
2.40. 0.02 milliohm
2.41. $1.37 \times 10^{-4}\Omega$
2.42. 1.55×10^{-14} teraohm
2.43. 0.035Ω
2.44. 0.001K
2.45. 1.5×10^{-7} megohm
2.46. 1×10^{-8} gigohm
2.47. 1×10^{-9} teraohm
2.48. 2692.6Ω
2.49. 270.5Ω
2.50. 32.17K
2.51. 3947.4Ω
2.52. 5870.8Ω
2.53. 57.35 degrees
2.54. 46.7 degrees
2.55. 13.26K
2.56. 0.1591 μF
2.57. 18,745 Hz
2.58. 45 degrees
2.59. 86.36 degrees
2.60. 1.374 radian
2.61. 150.79K
2.62. 918.9 Hz
2.63. 261.5 mH
2.64. 928.5Ω
2.65. 4649.6Ω
2.66. 32.15 degrees
2.67. 3.31 degrees
2.68. 29.488 kHz
2.69. 2200Ω
2.70. 29.59Ω
2.71. $R = 628\Omega$
$X = 136.2\Omega$
2.72. $R = 11.31\Omega$
$X = 11.31\Omega$
2.73. 1060Ω
2.74. 6183.7Ω
2.75. 155.6Ω
2.76. 85.1 degrees
2.77. 464.88K
2.78. 1.69 milliohm
2.79. 0.045 milliohm
2.80. 72.59 degrees
2.81. 1414Ω

2.82. 467.4Ω
2.83. 109.1Ω
2.84. 71.9Ω
2.85. 125Ω
2.86. 0.113 to 1
2.87. 100 to 1
2.88. 4-, 2-, 5-, 400-, 900-, and 2500 to 1, respectively
2.89. 150Ω
2.90. 1800Ω
2.91. 360Ω
2.92. 0.494 inch
2.93. xa = 8.42 ft
yb = 10.57 ft

2.94. 289.6Ω
2.95. 501Ω
2.96. 250.2Ω
2.97. 1.727 nepers
2.98. 26.06 dB
2.99. (a) −13.98 dB
(b) 1.609 nepers

2.100. $Z_1 = 447.2\Omega$
$Z_3 = 559\Omega$

2.101. $Z_1/2 = 223.6\Omega$
$Z_3 = 559\Omega$

2.102. $Z_1 = 941.9\Omega$
$Z_2 = 74.74\Omega$
$Z_3 = 78.29\Omega$

2.103. $Z_1/2 = 470.9\Omega$
$Z_2/2 = 37.37\Omega$
$Z_3 = 78.29\Omega$

2.104. $Z_1 = 1724\Omega$
$Z_2 = 251.3\Omega$
$Z_3 = 589.7\Omega$

2.105. $Z_1 = 1724\Omega$
$Z_2 = 251.3\Omega$
$Z_3/2 = 294.8\Omega$

2.106. 707.1Ω
2.107. 0.0146 mho
2.108. 33.3 milliohms
2.109. 10Ω
2.110. 99.62%
2.111. 0.53Ω
2.112. 314.2
2.113. 7854Ω
2.114. 3183Ω

Chapter 3

3.1. 310Ω
3.2. 1.97 milliohms
3.3. 0.25Ω
3.4. 2000Ω
3.5. 611Ω
3.6. 1.79Ω
3.7. 135.7Ω
3.8. 77.5Ω
3.9. 164.6K
3.10. 58.8K
3.11. 57.27Ω
3.12. 5.4Ω
3.13. 30.3Ω
3.14. 1875Ω
3.15. 75Ω
3.16. 1.02Ω
3.17. 100Ω
3.18. 5.67 times

3.19. 849 milliohms
3.20. 0.318Ω
3.21. 50.26Ω
3.22. 900Ω
3.23. 1591.5Ω
3.24. 24.87Ω
3.25. 909Ω
3.26. 62.5Ω
3.27. 227.3Ω
3.28. 72.2Ω
3.29. 630Ω
3.30. 2.67
3.31. 1.0
3.32. (a) $L_X = 1.234H$
(b) $R_X = 405.3\Omega$
(c) $Z_X = 3127.7\Omega$

3.33. (a) $L_X = 0.95$ mH
(b) $R_X = 11.9$ milliohms
(c) $Z_X = 5.97\Omega$.

Chapter 4

4.1. 15 μH
4.2. 70.7
4.3. 200 μH
4.4. 11.1%
4.5. 2.49 mH
4.6. No
4.7. 0.964H
4.8. 4000
4.9. 1.79 mH
4.10. 7160 μH
4.11. 0.866H
4.12. 0.067 μH
4.13. 21.05Ω
4.14. 61 mH
4.15. 0.0013H

Chapter 5

5.1. c
5.2. a
5.3. b
5.4. d
5.5. d
5.6. a
5.7. e
5.8. b
5.9. b
5.10. a
5.11. c
5.12. e
5.13. a
5.14. b
5.15. e
5.16. c
5.17. a
5.18. b
5.19. a
5.20. c
5.21. e
5.22. d
5.23. b
5.24. a
5.25. e
5.26. a
5.27. b
5.28. a
5.29. a
5.30. c
5.31. b
5.32. d
5.33. c
5.34. e

5.35. a
5.36. c
5.37. d
5.38. c
5.39. a
5.40. c
5.41. a
5.42. b
5.43. e
5.44. d
5.45. TRUE
5.46. FALSE
5.47. TRUE
5.48. TRUE
5.49. FALSE
5.50. TRUE
5.51. TRUE
5.52. FALSE
5.53. FALSE
5.54. TRUE
5.55. FALSE
5.56. FALSE
5.57. FALSE
5.58. TRUE
5.59. TRUE
5.60. FALSE
5.61. TRUE
5.62. FALSE
5.63. FALSE
5.64. TRUE
5.65. TRUE
5.66. FALSE
5.67. FALSE
5.68. TRUE
5.69. TRUE
5.70. TRUE
5.71. FALSE
5.72. FALSE
5.73. FALSE
5.74. TRUE
5.75. TRUE
5.76. FALSE
5.77. TRUE
5.78. FALSE
5.79. TRUE
5.80. TRUE
5.81. TRUE
5.82. FALSE

Index

Index

L

M